AF248715

Frontispiece. The H. H. Bloomer Medal. This silver medal is part of the H. H. Bloomer Award, which was established in 1963 by the Council of the Linnean Society of London from a legacy given by the late Harry Howard Bloomer, F.L.S. The award is given at intervals to an amateur naturalist, who has made an important contribution to biological knowledge. The obverse side of the medal depicts the young Linnaeus with "my flower" *Linnaea borealis* (p. 112) and the reverse bears the words H. H. Bloomer Award and the name of the recipient. It was with particular pleasure that the award for 1972 was made to Dr Marie Åsberg, who is a practising psychiatrist. (See *Biol. J. Linn. Soc., 4*: 339 (1972) for the full citation.)

Linnaeus's
Öland and Gotland Journey
1741

Translated from the Swedish edition 1745
by MARIE ÅSBERG and WILLIAM T. STEARN

With an Introduction by
William T. Stearn

Reprinted from the Biological Journal of the Linnean Society Volume 5, 1973

Published for the Linnean Society of London by Academic Press

ACADEMIC PRESS INC. (LONDON) LIMITED
24/28 Oval Road
London NW1
(Registered Office)
(Registered number 598514)

US edition published by
ACADEMIC PRESS INC.
111 Fifth Avenue
New York
New York 10003

LCCCN 73 17636
ISBN 0 12 064750 8

Printed in Great Britain by
The Whitefriars Press Ltd., London and Tonbridge

Biol. J. Linn. Soc., 5: 1-107

March 1973

Linnaeus's Öland and Gotland Journey 1741

Translated from the Swedish edition 1745

by MARIE ÅSBERG*

Department of Psychiatry, Karolinska Sjukhuset, 104 01 Stockholm, Sweden

and WILLIAM T. STEARN, F.L.S.

Department of Botany, British Museum (Natural History), London SW7 5BD

With an Introduction

by

William T. Stearn

CONTENTS

INTRODUCTION

IMPORTANCE OF LINNAEUS'S *ÖLÄNDSKA OCH GOTHLÄNDSKA RESA*

Carl Linnaeus's *Öländska och Gothländska Resa på Riksens Högloflige Ständers Befallning förrättad Ahr 1741 med Anmärkingar uti Oeconomien, Natural-Historien, Antiquiteter &c* (Stockholm & Uppsala, 1745) is often cited in his *Species Plantarum* (1753), *Flora Suecica,* 2nd ed. (1755) and *Fauna Suecica,* 2nd ed. (1761) as *"It. oel."* and *"It. gotl."*, but outside Sweden it is one of the least known and least consulted of Linnaean works. It gives an account of his expedition in the summer of 1741 to the Baltic islands of Öland and Gotland, with observations on their economy, products, natural history,

* Recipient of the H. H. Bloomer Award for 1972 (see citation in *Biol. J. Linn. Soc., 4:* 339).

antiquities, etc. This wide range of subject material reflects Linnaeus's eager and embracing curiosity and his belief that whatever merited his attention would likewise interest his contemporaries and would deserve publication, but the book still has a high intrinsic value. On the journey, because the two islands differ climatologically and geologically from the mainland of Sweden and shelter many organisms there unknown or rare, Linnaeus recorded several which were new to him then and to which he later gave binomials associated with diagnoses based primarily upon Öland and Gotland specimens. The *Öländska och Gothländska Resa* thus provides information relevant to the typification of some Linnaean specific binomials for plants, molluscs, insects and birds, of which Gotland or Öland proves to be the restricted type-locality. This travel book has also a special historical significance in relation to the development and adoption of Linnaean binomial nomenclature for species. Here Linnaeus first used purely designatory two-word specific names, such as *Veronica spicata, V. officinalis, V. scutellata, V. beccabunga, V. hederaefolia,* concurrently with descriptive and diagnostic several-word specific names such as *Veronica humilis erecta montana, flore parvo caeruleo.* Thus it foreshadows the later use of consistent binomial nomenclature for species in Linnaeus's *Species Plantarum* (1753) and *Systema Naturae* (1758) which has made these much larger works the internationally accepted starting points for the modern scientific nomenclature of plants and animals. Its nomenclatural importance and general interest are much greater than naturalists have hitherto appreciated.

The first edition, mostly in Gothic type and with now obsolete Swedish spelling, is admittedly unattractive typographically, but the Wahlström & Widstrand edition of 1962, beautifully illustrated by Gunnar Brusewitz, edited, modernized as regards Swedish spelling and helpfully annotated by C. O. von Sydow, is an enticing book. Bertil Gullander has produced in his *Linné på Öland* (1970) a profusely illustrated and annotated selection of extracts from both Linnaeus's manuscript and printed accounts of his Öland journey and in his *Linné på Gotland* (1971) a similar Gotland anthology. Nevertheless, although the Swedish text has been several times reprinted, e.g. in 1890, 1907, 1929, 1957, 1962, 1970 and 1971, and was translated into German by J. C. D. Schreber in 1764, it has never been put as a whole into English; only a few short extracts have appeared in biographies of Linnaeus, e.g. by Gourlie (1953) and Blunt & Stearn (1971). The present translation by Maria Åsberg and the writer thus makes it generally available and should assist in the typification of some Linnaean names as well as providing a first-hand account of a pioneering investigation of Swedish natural history.

ECONOMIC BACKGROUND

The background of Linnaeus's official travels in Sweden, to Öland and Gotland in 1741, to Västergötland (Westgothland) in 1746 and to Skåne in 1749, at the request of the Swedish Estates of the Realm, was an economic one, as Heckscher (1942) has emphasized. Linnaeus grew up in a Sweden impoverished by the wars of Charles XII but intent upon economic recovery. From 1739 to 1765 the Hats party, led for many years by Linnaeus's patron, the politician Count Carl Gustav Tessin, ruled the country and adopted a strongly mercantilistic policy intended to foster national self-sufficiency and home production by using wherever possible Swedish rather than imported raw materials. These national resources required survey, and Linnaeus zealously promoted their investigation. Thus the Swedish Royal Academy of Science, of which he was a founder member, had as its original aims the promotion of economics, trade, useful arts and manufactures as well as natural history and mathematics. Linnaeus himself was completely in sympathy with this utilitarian emphasis. His thrifty Småland upbringing and early poverty, his patriotism and the general attitude of his times made him keenly alert to possible benefits of science. His journeys were intended to be not so much natural history forays as preliminary assessments of Swedish natural resources and potential commercial possibilities, even though they did not really produce much of value in this way. Consequently his travel journals cover a much wider field of interest than botany and zoology and are written in a simple direct style.

JOURNEY TO ÖLAND

In 1741 Öland and Gotland together offered a rich and almost scientifically virgin area for investigation, although Rudbeck the Elder had listed as long ago as 1685 such distinctive Öland plants as *Helianthemum oelandicum* and *Euphorbia palustris.* Accordingly on 15 May 1741 Old Style (=26 May 1741 New Style), shortly after achieving his long-desired appointment as an Uppsala professor, Linnaeus set out from Stockholm, accompanied by six "handsome and intelligent youths", dividing between them the subjects of research. Linnaeus, the expedition leader, was just 34 years old. They travelled, by way of Svalbro, Nyköping and Norrköping, south to Växjo in Småland, where Linnaeus had spent his youth, and then south-east to the port of Kalmar, which they had reached on 28 May. The weather was cold and the sea was stormy, but its danger and discomfort were less daunting than the risk of catching fever in Kalmar; after some days of anxious waiting they took the first available boat to Öland, landing on 1 June 1741 at Färjestaden near Borgholm and exploring the island to 21 June. Öland is about 85 miles (137 km) long and at most 10 miles (16 km) broad. On this low-lying, almost flat limestone plateau, which has an arid maritime climate, grow some 1050 species of vascular plants (cf. Sterner, 1938; Sterner & Curry-Lindahl, 1955). The *alvar*-karst, with the limestone riven by deep fissures, in some ways resembles the limestone pavements of the Burren region of western Ireland and, like this, has depressions, which are water-covered in winter but often dried out in summer, where *Potentilla fruticosa* grows abundantly. The distinctive flora of the island includes a few species not found elsewhere in Sweden, among them *Artemisia oelandica* (*A. laciniata* auct. Scand.), *Helianthemum oelandicum, Plantago tenuiflora* and *Ranunculus illyricus,* and also many species which otherwise occur in Sweden only on the island of Gotland, among them *Adonis vernalis, Anacamptis pyramidalis, Anemone sylvestris, Coronilla emerus, Fumana procumbens, Globularia vulgaris* and *Potentilla fruticosa.* The island has 27 species of *Orchidaceae.* Linnaeus and his companions thus found Öland quite unlike any other Swedish province. Flora, as Linnaeus enthusiastically wrote, had provided her guests with a magnificent feast.

From Borgholm they travelled southward along the west side of the island to

Ottenby in the extreme south, then northward along the east side of the island. On 15 June they made an excursion to the island of Blåkulla or Jungfrun (cf. Du Rietz, 1925; Ottosson, 1965), situated in the north of the Kalmar Sound about 5½ miles (9 km) from Öland. This is an uninhabited rounded wooded granitic island about 3300 ft (1000 m) by 2600 ft (800 m), rising to 280 ft (86 m), now a Swedish national park and nature reserve. Linnaeus recorded 87 species of flowering plants, two ferns, one moss and three lichens. Not until Johan Erikson visited it in 1902, 1903 (cf. Erikson, 1904) and 1904 was its vegetation studied again.

JOURNEY TO GOTLAND

On 21 June 1741 Linnaeus and his party sailed from northern Öland for Gotland, landing at Visby on 22 June. Gotland, about 77 miles (125 km) long, and at most 32 miles (52 km) broad, is larger than Öland, has a more varied topography and a more northern position, but possesses approximately the same number of species. Their floras have much in common but are not identical. Thus, as noted above, Öland possesses species which do not grow on Gotland. The latter island, however, has several species not found on Öland, among them *Arenaria gothica, Bartsia alpina, Calamagrostis varia, Cephalanthera rubra, Inula ensifolia, Pinguicula alpina, Pulsatilla patens* and *Tofieldia calyculata* (cf. Pettersson, 1958, 1965, 1966). It thus provided Linnaeus with further botanical surprises.

From Visby they travelled northward along the west coast to Lummelund, Hangvar, Fleringe and Hau in the extreme north, then crossed the Färö sound to the Färö island and back again, and went southward from Bunge along the east coast to Gothum, Torsborgen, När, and ultimately to Hoburg in the extreme south, then northward along the west coast to Visby. On 13 July they made an excursion to the two Karlsöarne, two small islands Lilla Karlsö and Stora Karlsö, off the south-west coast of Gotland, which are called *Carolina* in Linnaeus's *Flora Suecica,* and stayed overnight in a fisherman's hut on Stora Karlsö. The most notable discovery here was of the rarest of all Scandinavian plants, *Lactuca quercina* found on Lilla Karlsö and recorded in the *Species Plantarum, 2* : 795 (1753) as "Habitat in Insula Carolina Balthici"; it occurs nowhere else in Scandinavia. Other rarities were *Gypsophila fastigiata* and *Artemisia rupestris.* On 25 July they sailed from Visby for Kalmar. The sea was again rough but, apart from sea-sickness, they reached Kalmar on 27 July without mishap. Here the party broke up and Linnaeus made his way to Stenbrohult in Småland, staying there four days with his father, his brother, sister and relatives.

Three of his companions now joined him here and they then made their way back to Stockholm by way of Jönköping and Vadstena. It had been a highly interesting trip and Linnaeus was well satisfied by it: "Altogether by this journey so far as natural history is concerned, I have discovered more than one could have thought possible. For botany alone I have found one hundred species of plants not known before in Sweden." He lost no time in communicating these discoveries to the Royal Academy of Sciences, for the *Kongl. Swenska Wetenskaps Academiens Handlingar* of July-Sept. 1741 (pp. 179-210) contains his account of a hundred plants found in Gotland,

Öland and Småland investigated on this journey; he also added references to Öland and Gotland species to his paper classifying the *Orchidaceae*, published in 1744. On 17 October 1741 Linnaeus delivered his inaugural address as an Uppsala professor, an *Oratio de Necessitate Peregrinationum intra Patriam*, in which, dwelling upon the need to travel in one's own country, he naturally instanced the value of his own travels, "made not without great fatigue of body and mind", taking some examples from this Öland and Gotland journey.

PREPARATION OF THE *RESA*

During the next few years Linnaeus was so busy with the duties of his professorship and with the preparation of his *Flora Suecica* (1745) that not until 1745 was he able to print his *Öländska och Gothländska Resa* describing this journey. It is in the form of a day-to-day rather full diary, with lists of plants, transcriptions of runic inscriptions, diagnoses and descriptions in mingled Latin and Swedish of plants and animals, and much miscellaneous information. The printed version diverges in many details from the manuscript diary written on the journey. This is in the library of the Linnean Society of London, with a photocopy in the Uppsala University library; some extracts from it have been published by Gullander (1970, 1971). Evidently at the end of the day's activities Linnaeus dictated an account to one of his companions acting as secretary, an account which incorporated their notes made in the field; three styles of handwriting and spelling are evident in the manuscript, the major part being attributed to Hans Jacob Gahn. Comparison by Sigurd Fries (1967) of the original and the printed version shows that, when rewriting the whole for printing, Linnaeus omitted material, especially that criticizing persons, and also made alterations for stylistic reasons. According to Fries, "Linnaeus has obviously struggled for a more correct, less colloquial style than that of the diaries".

INTRODUCTION OF BINOMIAL NOMENCLATURE

For species already well-known the nomenclature in the text of the *Öländska och Gothländska Resa* follows no consistent system. In a list of plants found at a given place Linnaeus used mononomials, binomials and polynomials side by side, e.g. *Marrubium, Agrimonia, Tithymalus helioscopius, Ranunculus echinatus, Hypericum caule ancipiti, Linum catharticum, Lotus corniculata, Alsine gramineo folio minor.* For new or little known species, however, he provided descriptive and diagnostic polynomials drafted in accordance with the precepts of his *Critica botanica* (1737), which he also used in his *Flora Suecica* (1745), e.g. *Lepidium foliis pinnatis integerrimis, petalis calice minoribus; Anthericum foliis ensiformibus, perianthiis trilobis, filamentis glabris; Lotus leguminibus solitariis membranaceo-quadrangularibus, foliolis floralibus lanceolatis; Lactuca foliis pinnato-sinuatis denticulatis acutis subtus laevibus, caule glabro.* Such names were not jotted down in the field, like those of his floristic lists, but were obviously constructed when Linnaeus worked over his Öland and Gotland material for inclusion in his comprehensive *Flora Suecica.* This and the *Öländska och Gothländska Resa* must be consulted together, as the latter supplies details sometimes omitted from the former. The nomenclature of the

Flora Suecica is of "pre-Linnaean" polynomial character throughout, like Linnaeus's earlier *Flora Lapponica* (1737) and *Hortus Cliffortianus* (1738); each specific entry has its own serial number, the numbering being continuous from 1 to 1140, thus making reference quick and convenient. Instead of a long polynomial, such as *Lithospermum seminibus laevibus, corollis vix calycem superantibus,* the entry *Lithospermum Fl. Suec.* 151 sufficed to indicate precisely which plant was meant. In the index to the *Öländska och Gothländska Resa,* however, Linnaeus adopted an intermediate system of naming. Here he listed the plants in the same sequence as in the *Flora Suecica,* in accordance with the classes of his "sexual system" of classification, i.e. under Monandria, Diandria, Triandria, etc. (cf. Blunt & Stearn, 1971 : 244), used the same generic name and number and then added a catchword or epithet. Thus the four species of *Lepidium, Anthericum, Lotus* and *Lactuca* named above were indexed as *Anthericum* 269 *calyciflorum, Lotus* 610 *marina, Lepidium* 535 *minimum, Lactuca* 646 *quercifolia.* These numbers concisely link the text of the *Öländska och Gothländska Resa* with the *Flora Suecica;* Linnaeus did not include the long diagnostic phrase-names of his text in the index. "Evidently, when indexing his travel-book, Linnaeus found the phrase-names inconveniently long, substituted the *Flora Suecica* number instead, disliked having a specific number alone, so added to it a single catchword easy to remember, thus creating the binomial system of nomenclature" (Stearn, 1957 : 68). This simple nomenclatural invention of far-reaching importance makes the *Öländska och Gothländska Resa* a landmark in biological history. Linnaeus's consistent use of binomial nomenclature for species, though grounded in the usage of common speech going back to remote antiquity, was not derived from the incidental and inconsistent use of binomials by Linnaeus's predecessors but was a deliberate innovation allied to Linnaeus's earlier use of binomial citations for books, e.g. *Clus. hisp., Clus. pan., Clus. hist.* and *Clus. exot.* for four works by Carolus Clusius (Charles de l'Ecluse), of which the full titles may be compared to the polynomial names of species, e.g. *Caroli Clusii Rariorum aliquot Stirpium per Hispanias observatarum Historia* (1576). Linnaeus did not intend such binomials to supplant his carefully elaborated polynomials but to be used concurrently with them, for they had different functions. To the lasting benefit of science he separated as incompatible their designatory and diagnostic functions. This matter has been discussed by Stearn (1959) and Heller (1964). Linnaean binomial nomenclature began as an indexer's paper-saving device.

Linnaeus next used binomial nomenclature for plant species in the dissertation *Pan Suecicus* (1749) and here he explained, as he had not done earlier, his new method of listing species: "If we take the *Flora Suecica,* Stockholm 1745, and for each herb, in order to save paper, we put the generic name, the number of the *Flora Suecica* and some epithet in place of the differential diagnosis, the matter is easily put in handy form." He used the same method in the dissertations *Gemmae Arborum* (1749), *Splachnum* (1750) and *Plantae esculentae Patriae* (1752) for a limited number of plants before applying it to the vegetable kingdom as a whole in his *Species Plantarum* (1753). A binomial could not exist independently; it is meaningless because of uncertain application unless it is definitely linked to a diagnosis or description. Without the *Flora Suecica* to provide a firm foundation, Linnaeus could not have introduced binomial nomenclature in the *Öländska och Gothländska Resa* or elsewhere for Swedish plants. The world-wide application of consistent binomial nomenclature to animals, first used for a limited number of them in his *Hospita Insectorum Flora* (1752) and *Museum Tessinianum* (1753) and Clerck's *Svenska Spindlar* (1757), had to await the completion of the 10th edition of his *Systema Naturae* (1758-59).

RESTRICTED TYPE-LOCALITIES OF PLANTS

Linnaeus's statements of range in the *Species Plantarum* (1753) and *Systema Naturae* (1738) are usually in such vague general terms, e.g. "Habitat in Europae paludibus", "Habitat in alpibus Helvetiae, Lapponiae, Sibiriae", as to be virtually useless if taken at their face value, but, by analysis of his sources of information, some at least of his statements can be made more precise or even narrowed to definite places (cf. Stearn, 1957 : 143-150). Examination of a Linnaean protologue usually reveals one visual element upon which his diagnostic polynomial was primarily based and which should be selected as the lectotype; the provenance of this, when it can be ascertained, may then indicate a restricted type-locality from which topotype material may possibly be obtained for cytological and other purposes. Thus when Linnaeus took unchanged into the *Species Plantarum* a diagnosis from his *Flora Suecica* which had been based on Öland or Gotland material, and especially when he directed particular attention to the account in his *Öländska och Gothländska Resa* by means of an asterisk, e.g. *It. gotl.*, 192*, quite certainly he accepted the Öland or Gotland plant as representing the species concerned; for such species Öland or Gotland can reasonably be taken as the restricted type-locality.

The relevance of Gotland or Öland material to the typification of names in the *Species Plantarum* can be illustrated, for example, by *Geranium lucidum* L., *Sp. Pl., 2* : 682 (1753) of which the protologue is as follows:

> 32. GERANIUM pedunculis bifloris, calycibus pyramidatis angulatis rugosis, foliis quinquelobis rotundatis [i]. *It. gotl.*, 228*. *Fl. Suec.*, 574. *Dalib. paris.*, 208. *Sauv. monsp.*, 207 [ii].
> Geranium lucidum saxatile. *Bauh. pin.*, 318. [iii].
> Geranium saxatile. *Thal. herc.*, 44, *t.* 5. [iv].
> Geranium rotundifolium saxatile montanum. *Col. ecphr.*, *1* : 138, *t.* 137. [v].
> *Habitat in* Europae *rupibus umbrosis* [vi].

This consists of three main parts, viz. the specific diagnosis [i] with its place of publication and acceptance [ii], synonyms in earlier literature [iii-v], its general distribution and habitat [vi]. The synonyms come from C. Bauhin, *Pinax* (1623), J. Thal, *Sylva Hercynia* (1588) and F. Colonna, *Minus cognitarum Stirpium Ekphrasis* (1616); those of Thal [iv] and Colonna [v] have illustrations which portray the habit. Under the diagnosis itself Linnaeus referred to his own *Öländska och Gothländska Resa* (1745), his *Flora Suecica* (1745), T. F. Dalibard, *Florae Parisiensis Prodromus* (1749) and Boissier de Sauvages, *Methodus Foliorum, seu Plantae Florae Monspeliensis* (1751). Collectively these references indicate the occurrence of the species in Sweden, France, Germany and Italy and thus give a little more precision to Linnaeus's generalized statement of range. Linnaeus himself knew the plant first hand from Swedish material. His crucial reference, specially indicated by an asterisk implying a good description or important information, is to "*It. gotl.*, 228", i.e. the *Öländska och Gothländska Resa,* 228 (1745). His account here, relating to plants found on 3 July 1741 at Torsborgen, Gotland, is as follows:

> "Geranium *pedunculis bifloris, calycibus pyramidatis angulatis rugosis, foliis quinquelobis rotundis,* en wacker Ört, den Thalius in *Hercyn., t.* 5 och Columin in *Ecphr., t.* 137 artigt afritat, men tilförna aldrig warit plåckad i Swerige, wäxte besynnerligen wid norra Sidan på de nedfalne Klippo uti skuggan af Thorsborgen. Roten går bort hwart åhr: Stielken (och Bladen besynnerligen inunder) röd, slätt och glatt. Bladen äro Niur-lika, glatta, försedde med Bladstaft, femdelta, och hwar flik tredelt. Blomskaften hålla twå Blommor. Blomfodren äro något upblåste, glatte, fem-Bladige, dock trekantige, hwar och en kant pa hwardera sidan med 3 uphögda twärstrimmor utmärkt. Kronbladen äro aldeles hela och kiött-färgade." [Geranium . . . , a beautiful herb, which Thalius in *Hercyn., t.* 5 and Columna in *Ecphr., t.* 137 skilfully portray, but which previously had never been gathered in Sweden, grew especially on the northern side of the fallen down rocks in the shade of Thorsborgen. The root dies every year; the stem (and the leaf-blade especially on the underside) red, even and smooth. The leaf-blades are kidney-shaped, smooth, provided with a stalk, 5-parted, and every segment 3-parted. The flower stalk holds 2 flowers. The calyces are somewhat swollen, smooth, 5-leaved, but nevertheless 3-angled, every angle marked on each side with 3 raised transverse stripes. The petals are quite entire and flesh-coloured.]

This Gotland material provided Linnaeus with his first opportunity to study from living plants a species known to him imperfectly from two illustrations in the literature and on it he based the diagnosis or phrase-name which he repeated in the *Species Plantarum*; it was thus the basic element in his concept of the species as defined here and later.

In the index to the *Resa* this is entered not under the diagnostic phrase-name *Geranium pedunculis bifloris* etc. but as "Geranium 574 *lucidum* 72, 228". The number "574" refers to the entry 574 in his *Flora Suecica,* 207 (1745) which is as follows:

> 574. GERANIUM pedunculis bifloris, calycibus pyramidatis angulatis rugosis, foliis quinquelobis rotundatis. *Act. stockh.,* 1741 : 200.
> Geranium lucidum saxatile. *Bauh. pin.,* 318.
> Geranium lucidum. *Bauh. hist., 3* : 481.
> Geranium saxatile. *Raj. hist.,* 1060. *Thal herc.,* 44, *t.* 5.
> Geranium rotundifolium saxatile montanum. *Col. ecphr., 1* : 138, *t.* 137.
> Habitat in *lateribus montibus Thorsburg Gotlandiae; Stenhufwud & Trollarestugan ad Glimrakra* Scaniae.

To the Gotland localities Linnaeus here added localities in Skåne but he also cited an earlier publication of his diagnostic phrase-name in *Act. stockh., 1741 : 200.* This refers to an entry in *Kongl. Swenska Wetensk. Acad. Handlingar* 1741 : ii, 200 (1741) published soon after Linnaeus's return from his Öland and Gotland journey in a paper enumerating a hundred new or interesting plants then discovered. It reads as follows:

> 64. GERANIUM pedunculis bifloris, calycibus pyramidatis angulatis glabris, foliis quinquelobis rotundatis.
> Geranium lucidum saxatile. *Bauh. pin.*, 318.
> Geranium lucidum. *Bauh. hist.*, *3* : 481.
> Geranium saxatile. *Raj. hist.*, 1060.
> Waxer på Torsborg på Gotland.
> Stielken är klar; Bladen merendels röda inunder; Bloman sirat med röda strimmor.

Thus, running through all this Linnaean literature from 1741 to 1753, which relates to *Geranium lucidum*, is its occurrence on Gotland, specifically at Torsborgen, the primary source of Linnaeus's own knowledge of the plant. Hence Gotland is to be accepted as the restricted type locality of *Geranium lucidum.*

The following is a list of the Linnaean species with Öland or Gotland as their restricted type locality; it has been based, as above, on correlation of entries in the *Öländska och Gothländska Resa* (1745), *Flora Suecica* (1745; 2nd ed. 1755), *Kongl. Swenska Wetensk. Acad. Handlingar* 1741 and *Species Plantarum* (1753).

Allium scorodoprasum L., *Sp. Pl.*, *1* : 297 (1753), *Fl. Suec.*, 2nd ed., 103, no. 278 (1755); Hultén, *Atlas,* 122, map 487 (1950, 1971).
"Habitat in Oelandia, Dania, Pannonia" (*Sp. Pl.*).
Restricted type-locality: Öland, where Linnaeus collected it on 4 June 1741 near Glömminge (cf. *Öländ. Resa,* 60; *Fl. Suec.*, 94, no. 266); frequent on Öland (cf. Sterner, 1958: 81).
Linnaean Herbarium sheet 419.12.

Anthericum calyculatum L., *Sp. Pl.*, *1* : 311 (1753), *Fl. Suec.*, 2nd ed., 107, no. 288 (1755).
=*Tofieldia calyculata* (L.) Wahlenb., *Fl. Lapp.*, 90 (1812); Hultén, *Atlas,* 120, map 475 (1950, 1971).
"Habitat in alpibus Helvetiae, Lapponiae, Sibiriae" (*Sp. Pl.*).
Restricted type-locality: Gotland, where Linnaeus collected it on 28 June 1741 near Hau (cf. *K. svenska VetenskAkad. Handl.*, 1741; ii, 191; *Öländ. Resa,* 194; *Fl. Suec.*, 95, no. 269; Stearn, 1947: 194; Stearn, 1957: 131).

Artemisia rupestris L., *Sp. Pl.*, *2* : 847 (1753), *Fl. Suec.*, 2nd ed., 285, no. 733 (1755); Hultén, *Atlas,* 443, map 1726 (1950, 1971).
"Habitat in Siberia, Oelandiae rupibus calcareis" (*Sp. Pl.*).
Restricted type-locality: Gotland, on the island Stora Karlsö, where Linnaeus collected it on 14 July 1741 (cf. *K. svenska VetenskAkad. Handl.*, 1741; II, 205; *Oländ. Resa,* 285, with fig.; *Fl. Suec.*, 240, no. 669).
Linnaean Herbarium sheet 988.25.

Astragalus campestris L., *Sp. Pl.*, *2* : 761 (1753), *Fl. Suec.*, 2nd ed., 257, no. 662 (1755).
=*Oxytropis campestris* (L.) DC., *Astragal.* 74 (1802); Hultén, *Atlas,* 294, map 1143 (1950, 1971).

"Habitat in Oelandia, Germania, Helvetia" (*Sp. Pl.*).
Restricted type-locality: Öland, where Linnaeus collected it on 2 June 1741 near Räpplinge (cf. *Öländ Resa,* 46; *Fl. Suec.*, 214, no. 593); common on the alvar limestone (cf. Sterner, 1938: 120, fig. 180).
Linnaean Herbarium sheet 926.51, labelled "30 campestris Oelandia".

Atriplex hastata L., *Sp. Pl., 2* : 1053 (1753), *Fl. Suec.*, 2nd ed., 364, no. 921 (1755).
=*A. calotheca* (Rafn) Fries, *Nov. Fl. Suec.*, 2nd ed. Mant., *3* : 164 (1842); Hultén, *Atlas,* 167, map 656 (1950, 1971).
"Habitat in Europa frigidiori" (*Sp. Pl.*).
Restricted type-locality: Öland, where Linnaeus collected it on 8 June 1741 near Ottenby (cf. *Öländ. Resa,* 88; *Fl. Suec.*, 301, no. 827).
Linnaean Herbarium sheet 1221.17.

For a discussion of the typification of the name *A. hastata* L. see Taschereau in *Can. J. Bot.*, *50* : 1585 (1972); the correct name for the species commonly known as *A. hastata* is, as pointed out by Taschereau, *A. triangularis* Willd.

Carduus acaulos L., *Sp. Pl., 2* : 1199 (1753), *Fl. Suec.*, 2nd ed., 281, no. 722 (1755).
=*Cirsium acaule* Scopoli, *Ann. hist.-nat.*, *2* : 62 (1769); Hultén, *Atlas,* 451, map 1758 (1950, 1971).
"Habitat in Europae pascuis apricis, depressis" (*Sp. Pl.*).
Restricted type-locality: Gotland, where Linnaeus collected it on 20 July 1741 near Vible (cf. *K. svenska VetenskAkad. Handl.*, 1741; ii: 204; *Öländ. Resa,* 297; *Fl. Suec.*, 236, no. 656). The names *Carduus acaulos* L. and *C. acaule* Scopoli were based on different types, as noted by Dandy in *Watsonia,* *7* : 167 (1969), the first from Gotland, the second from the Italian Tyrol, which are, however, now regarded as conspecific.
Linnaean Herbarium sheet 966.45.

Carex arenaria L., *Sp. Pl., 2* : 973 (1753), *Fl. Suec.*, 2nd ed., 325, no. 835 (1755); Hultén, *Atlas,* 86, map 341 (1950, 1971).
"Habitat in Europae arena, praesertim mobili" (*Sp. Pl.*).
Restricted type-locality: Öland, where Linnaeus collected it on 18 June 1741 near Grankull (cf. *Öländ. Resa,* 139; *Fl. Suec.*, 270, no. 749); frequent on sandy soils (cf. Sterner 1938: 75, fig. 65).
Linnaean Herbarium sheet 1100.9, labelled "arenaria 4 Scania".

Cistus oelandicus L., *Sp. Pl., 1* : 526 (1753), *Fl. Suec.*, 2nd ed., 184, no. 473 (1755).
=*Helianthemum oelandicum* (L.) DC. in Lam. & DC., *Fl. Franç.*, 3rd ed., *4* : 817 (1805); Hultén, *Atlas,* 320, map 1243 (1950, 1971).
"Habitat in rupibus apricis Oelandiae" (*Sp. Pl.*).
Type-locality: Öland, where Linnaeus collected it on 5 June 1741 near Resmo (cf. *K. svenska VetenskAkad. Handl.*, 1741: ii, 194; *Öländ. Resa,* 70; *Fl. Suec.*, 158, no. 434).
Linnaean Herbarium sheet 689.40, labelled "14 oelandicus Oeland".

Geranium lucidum L., *Sp. Pl., 2* : 682 (1753), *Fl. Suec.*, 2nd ed., 241, no. 620 (1755); Hultén, *Atlas,* 304, map 1181 (1950, 1971).
"Habitat in Europae rupibus umbrosis" (*Sp. Pl.*).
Restricted type-locality: Gotland, where Linnaeus collected it on 3 July 1741 at Torsborgen (cf. *K. svenska VetenskAkad. Handl.*, 1741; ii, 200; *Öländ. Resa,* 228; *Fl. Suec.*, 207, no. 574); see discussion above.
Linnaean Herbarium sheet 858.72.

Globularia vulgaris L., *Sp. Pl., 1* : 96 (1753); *Fl. Suec.*, 2nd ed., 41, no. 116 (1755); Hultén, *Atlas,* 413, no. 1610 (1950, 1971).
"Habitat in Europae apricis duris" (*Sp. Pl.*).
Restricted type-locality: Öland, where Linnaeus collected it on 4 June 1741 near Resmo (cf. *K. svenska VetenskAkad. Handl.*, 1741: 187; *Öländ Resa,* 65; *Fl. Suec.*, 39, no. 109); infrequent but most abundant in the south (cf. Sterner, 1938: 151, fig. 253).
Linnaean Herbarium sheet 116.2.

Lactuca quercina L., *Sp. Pl., 2* : 795 (1753), *Fl. Suec.*, 2nd ed., 270, no. 691 (1755); Hultén, *Atlas,* 465, map 1812 (1950, 1971).
"Habitat in Insula Carolina Balthici" (*Sp. Pl.*).
Type-locality: Gotland, on the island Lilla Karlsö, where Linnaeus collected it on 14 July 1741 (cf. *K. svenska VetenskAkad. Handl.*, 1741; ii, 203; *Öländ. Resa,* 289; *Fl. Suec.*, 233 no. 646).
Linnaean Herbarium sheet 950.1.

Lepidium petraeum L., *Sp. Pl., 2* : 644 (1753), *Fl. Suec.*, 2nd ed., 225, no. 573 (1755).
=*Hutchinsia petraea* (L.) R. Br. in Aiton, *Hort. Kew.*, 2nd ed., *4* : 82 (1812).
=*Hornungia petraea* (L.) Reichenb., *Deutschl. Fl., 1* : 33 (1837); Hultén, *Atlas,* 226, no. 881 (1950, 1971).
"Habitat in Lapidosis Oelandiae, Angliae" (*Sp. Pl.*).
Restricted type-locality: Öland, where Linnaeus collected it on 2 June 1741 near Bornholm (cf. *K. svenska VetenskAkad. Handl.*, 1741; ii, 198; *Öländ. Resa,* 52; *Fl. Suec.* 195 no. 535); very frequent on Öland (cf. Sterner, 1938: 107).
Linnaean Herbarium sheet 824.7.

Lotus maritimus L., *Sp. Pl., 2* : 773 (1753); *Fl. Suec.*, 2nd ed., 262 no. 676 (1755).
=*Tetragonolobus maritimus* (L.) Roth, *Tent. Fl. Germ., 1* : 323 (1788); Hultén, *Atlas,* 292, map 1135 (1950, 1971).
"Habitat in Europae maritimis" (*Sp. Pl.*).
Restricted type-locality: Öland, where Linnaeus collected it on 18 June 1741 near Grankulla (cf. *K. svenska VetenskAkad. Handl.*, 1741; ii, 201; *Öländ. Resa,* 141; *Fl. Suec.*, no. 610).
Linnaean Herbarium sheet 931.1.

Melica ciliata L., *Sp. Pl., 1* : 66 (1753); *Fl. Suec.*, 2nd ed., 26, no. 77 (1755); Hultén, *Atlas,* 50, map 197 (1950, 1971).
"Habitat in Europae collibus sterilibus saxosis" (*Sp. Pl.*).

Restricted type-locality: Gotland, where Linnaeus collected it on 20 July 1741 near Vible (cf. *K. svenska VetenskAkad. Handl.*, 1741: ii, 183: *Öländ. Resa,* 297; *Fl. Suec.*, 220, no. 56).
Linnaean Herbarium sheet 86.1.

Pyrola minor L., *Sp. Pl., 1* : 396 (1753), *Fl. Suec.*, 2nd ed., 139, no. 361 (1755); Hultén, *Atlas,* 349, map 1362 (1950, 1971).
"In Europa frigidiore" (*Sp. Pl.*).
Restricted type-locality: Gotland, where Linnaeus collected it on 29 June 1741 in the extreme north on Färö (cf. *K. svenska VetenskAkad. Handl.*, 1741; ii, 192; *Öländ. Resa,* 206; *Fl. Suec.*, 121 no. 331).
Linnaean Herbarium sheet 568.3.

Schoenus mariscus L., *Sp. Pl., 1* : 42 (1753), *Fl. Suec.*, 2nd ed., 13, no. 38 (1755).
=*Cladium mariscus* (L.) Pohl, *Tent. Fl. Bohem., 1* : 32 (1809); Hultén, *Atlas* 78, map 311 (1950, 1971).
"Habitat in Europae paludibus" (*Sp. Pl.*).
Restricted type-locality: Gotland, where Linnaeus collected it on 25 June 1741 at Lummelund (cf. *Öländ. Resa,* 170; *Fl. Suec.*, 13, no. 35).

Schoenus nigricans L., *Sp. Pl., 1* : 43 (1753), *Fl. Suec.*, 2nd ed., 14, no. 39 (1755); Hultén, *Atlas* 78, map 309 (1950, 1971).
"Habitat in Europae paludibus" (*Sp. Pl.*).
Restricted type-locality: Gotland, where Linnaeus collected it on 4 July 1741 near Gamelgarn (cf. *Öländ. Resa,* 234; *Fl. Suec.*, 13, no. 36).
Linnaean Herbarium sheet 68.6, labelled "4 Gotland".

Scutellaria hastifolia L., *Sp. Pl., 2* : 599 (1753), *Fl. Suec.*, 2nd ed., 110, no. 539 (1755); Hultén, *Atlas* 381, map 1481 (1950, 1971).
"Habitat ad littora Sueciae rarius" (*Sp. Pl.*).
Restricted type-locality: Gotland, where Linnaeus collected it on 30 June 1741 near Bunge (cf. *K. svenska VetenskAkad. Handl.*, 1741; ii, 197; *Öländ. Resa,* 212; *Fl. Suec.*, 181, no. 500).

Trifolium procumbens L., *Sp. Pl., 2* : 772 (1753), *Fl. Suec.*, 2nd ed., 261, no. 673 (1755).
=*T. campestre* Schreber in Sturm, *Deutschl. Fl., 1*: Heft. 16 : 7 (1804).
"Habitat in Europae campestribus" (*Sp. Pl.*).
Restricted type-locality: Gotland, where Linnaeus collected it on 9 July 1741 near Hoburg (cf. *K. svenska VetensAkad. Handl.*, 1741; ii, 202; *Öländ. Resa,* 257; *Fl. Suec.*, 223, no. 618). Owing to its diverse applications by authors the name *T. procumbens* L. has been rejected, as *nomen ambiguum,* and the name *T. campestre* used for the present species; its distribution is mapped under the name *T. campestre* in Hultén, *Atlas* 288, map 1120 (1950, 1971).
Linnaean Herbarium sheet 930.62, labelled "38 Gotl.".

RESTRICTED TYPE-LOCALITIES OF ANIMALS

Correlation of some zoological entries in the *Öländska och Gothländska Resa* (1745) with those in the *Fauna Suecica* (1746; 2nd ed. 1761) and *Systema*

Naturae, 10th ed., *1* (1758) reveals that for a number of animals, as for plants (see above), Öland or Gotland can or must be taken as the restricted type-locality. A few examples should suffice to indicate the need for more detailed enquiry by zoological systematists.

On 20 July 1741 Linnaeus collected at Vible near Visby, Gotland, a species, probably of *Glomeris,* which he named *Oniscus cauda obtusa integerrima* in the *Resa,* 298 (1745) and listed, under the same diagnostic name, in his *Fauna Suecica,* 360, no. 1256 (1746) as "Habitat in pratis Gotlandiae". In the *Syst. Nat.,* 10th ed., *1* : 637 (1758) he provided this with the binomial *Oniscus armadillo,* referring back to *Fn. Svec.,* 1256; *It. gotl.,* 298 but giving its distribution vaguely as "Habitat in Europa sub lapidis". In the *Fauna Suecica,* 2nd ed., 500, no. 2059 (1761) Linnaeus retained the name *Oniscus armadillo.* The neighbourhood of Visby, Gotland is accordingly the restricted type-locality from which topotype material should be collected.

Cochlea balthica was so named by Linnaeus in *Syst. Nat.,* 10th ed., *1* : 775 (1758), the habitat being given as "ad M. Balthici littora" but *Fn. svec.,* 1316, *It. gotl.,* 261 being cited. Reference to the *Öländ. Resa,* 261 (1746), where it is entered as *Cochlea testa pellucida, anfractibus quatuor rictu ovato amplo, superficie rugis elevatis,* as also to *Fauna Suecica,* 376, no. 1316 (1746); 2nd ed., 532, no. 2193 (1761), shows that the type-locality is on the coast of Gotland, where Linnaeus collected it near Hoburg on 9 July 1741.

Arenaria interpres (L.), the species of wading bird called the "turnstone" in English and "roskarl" in Swedish, nests in both northern Europe and North America, being represented on each continent by a different race. In the *Syst. Nat.,* 10th ed., *1* : 148 (1938), where Linnaeus named it *Tringa interpres,* he gave its range as "in Europa & America septentrionali". He based his American record on the illustration of the "Turn-stone, from Hudson's Bay" in George Edwards, *Natural History of Birds, 3* : t.141 (1750), named *Morinellus Canadensis* in Edwards's index. Linnaeus identified this as belonging to the same species as the bird described at length in his *Fauna Suecica,* 56, no. 154 (1746) as *Tringa nigro albo ferrugineoque variegata, pectore abdomineque albo,* said to be called "Tolck" by the people of Gotland and found in "Gotlandiae insulis Heligholmen & Clasen". The entry in his Gotland diary for 1 July 1741 (*Öländ. Resa,* 217) describes the "Tolk (*Tringa nigro albo ferrugineque variegata, pectore abdomineque albo*)" as a bird which he had not seen anywhere but at Klasen and had not found recorded by any author, although it was in fact known to Willughby & Ray, being a migrant on British shores. Gotland is accordingly the restricted type-locality for *Arenaria interpres interpres,* the European race, as distinct from the American *A. interpres morinella* (L.). The Swedish word "tolk" means "interpreter" and Linnaeus accordingly gave the species the epithet (trivial name) *interpres.* It would seem, however, that the name "tolk" more correctly belongs to the redshank (*Tringa totanus* L.), which warns other birds of danger by its alarm cry "teuk".

Aparia crataegi (*Papilio crataegi* L., *Syst. Nat.,* 10th ed., *1* : 467; 1758), collected by Linnaeus near Hangvar, Gotland on 26 June 1741 and named *Papilio hexapus : alis erectis, rotundatis, albis; venis nigris* (*Öländ. Resa,* 182); *Galeruca tanaceti* (*Chrysomela tanaceti* L., *Syst. Nat.,* 10th ed., *1* : 369; 1758), collected by Linnaeus near Bursviken, Gotland, on 10 July 1741 and named *Chrysomela atra, punctis excavatis contiguis* (*Resa,* 270); and *Staphylinus littoreus* L., *Syst. Nat.*, 10th ed., *1* : 422 (1758), collected by Linnaeus at Martebo, Gotland on 25 June 1741 and named *Staphylinus niger, elytris antice griseis, pedibus rufis*, are examples of other species having Gotland as their restricted type-locality.

Malachius bipustulatus (*Cantharis bipustulata* L., *Syst. Nat.*, 10th ed., *1* : 402; 1758), collected by Linnaeus near Gaxa, Öland on 15 June 1741 and named *Cantharis aeneo-viridis, elytris apice rubris* (*Öland. Resa*, 126), and *Halyzia oblongo-guttata* (*Coccinella oblongo-guttata* L., *Syst. Nat.*, 10th ed., *1* : 367; 1758), collected by Linnaeus in the north of Öland near Bryum on 18 June 1741 and named *Coccionella coleoptris rubris lineis quatuor albis longitudinalibus* (*Resa*, 148), are examples of species having Öland as their restricted type-locality.

There are undoubtedly many others.

RUNIC INSCRIPTIONS

Öland and Gotland abound in monumental stones with engraved runic inscriptions. Linnaeus copied these on both islands, usually having first to strip off a covering of moss. For this task of transcription he had little experience and little time; moreover weathering over the centuries had made letters indistinct or had wholly obliterated them. He accordingly wished his copies to be regarded as provisional guides to the existence of such inscriptions at particular places rather than as finished scholarly transcriptions. His manuscript in the Linnean Society's library contains some runic copies made on Gotland, which have been published in Gullander's *Linné på Gotlånd* (1970), but Linnaeus himself published only those made on Öland, because Bishop Georg Wallin in his *Runographia Gothlandica* (1743-1744) had dealt with those on Gotland. Since then a number of these have disappeared but the extant inscriptions have been carefully studied by Swedish scholars and published in *Sveriges Runinskrifter* (Stockholm, 1900 et seq.) by the Royal Swedish Academy of Letters, History and Antiquities (Kungl. Viterhets Historie och Antikvitets Akademien); those of Öland are reproduced in vol. 1 (*Ölands Runinskrifter*). Comparison of Linnaeus's copies in the *Öländska Resa* with those of *Ölands Runinskrifter* shows that he misread a number of letters, which is not surprising, and his copies would be difficult to translate but for this later work. Dr Helmer Gustavson of the Riksantikvarieämbetet, Stockholm, has kindly provided both transliterations and translations of these. They are mostly simple monuments to the dead, such as "Tore and Torsten and Torfast, those brothers raised this stone in honour of Gunnfun their father. May God help his soul." Only occasionally do they give hints of events, such as "Halvboren sits in Gardarike", i.e. Russia, and "Unn was revenged by Olof" his son.

TRADITIONAL SWEDISH MEASURES USED BY LINNAEUS

As the metric system was not introduced until long after Linnaeus's death, he naturally used the traditional Swedish measures of his time, which were as follows:

1 *mihl (mil)* = 1 Swedish mile = 10.7 km = approx. 6⅔ English miles!
1 *famn* = 1 fathom = 1.8 m = approx. 6 feet.

1 *aln* = 1 ell = 60 cm = approx. 24 inches (2 feet).
1 *fot* = 1 foot = 30 cm = approx. 12 inches (1 foot).
1 *spann* = 1 span = 20 cm = approx. 8 inches.
1 *quarter (kvarter)* = 1 quarter-ell = 15 cm = approx. 6 inches.
1 *tum* = 1 inch = 2.5 cm = approx. 1 inch.

These measures were mostly based on the *aln* or ell; thus the *famn* was equivalent to 3 alnar; the *quarter* or *kvarter* was $\frac{1}{4}$ aln; the *åttendondel* was $\frac{1}{8}$ aln. The "mile", although originally based on the Roman *milia* of 1000 paces, roughly 1480 m, varied from country to country, that of Sweden being among the longest in Europe and equivalent to 1800 alnar, i.e. 10,688 m or roughly 10.7 km (see Knut Birkeland, *Mått mål vikt;* 1971). The English mile is 1609 m or 1.6 km. Thus the distances travelled by Linnaeus and recorded in Swedish miles are more than six times greater than they appear to be, a difference which has unpleasantly surprised some British travellers in Sweden innocently believing English and Swedish miles to be the same.

Linnaeus also used the more vague expression *byszeskott (bysseskott),* meaning "a gunshot", which obviously depends for its interpretation upon the kind of gun in mind. According to Dr C. J. Clemendson, Surgeon General of the Swedish Armed Forces, and Dr Olle Cederlöf, Director of the Army Museum, Stockholm, who have kindly enquired into this matter for me in consultation with some specialists on old Swedish measures and old Swedish hunting, Linnaeus's *byszeskott* referred to the range of a shot from a hunting gun. Thus A. Nordholm in his *Jämtlands Djur-Fänge* (1749) remarks that wild swans and wild geese are so watchful that one can get no closer to them than 300 paces, a distance at which a gunshot cannot hit them ("de äro sa uppmärksamme, att man näplig kan komma den när pa 300 steg, hvars längd intet bösse-skott mägtan träffa rätta fram"). The maximum range of a hunting gun in Linnaeus's time is supposed to have been about 225 m (246 yards). Obviously, although a *byszeskott* can be taken as equivalent to about 225 m, Linnaeus judged such a distance by eye without any intent at great exactness. An English writer of this period would probably have translated it as "furlong", i.e. 201 m (220 yards).

SOME SOURCES OF INFORMATION

BLUNT, W. & STEARN, W. T., 1971. *The compleat naturalist: a life of Linnaeus.* London: Collins.

DU RIETZ, G. E., 1925. Die Hauptzüge der Vegetation der Insel Jungfrun. *Svensk. bot. Tidskr., 19:* 321-346.

ERIKSON, J., 1904. En studie öfver Jungfruns fanerogam-vegetation. *Ark. Bot., 2,* no. 3.

FRIES, S., 1967. Linnés resedagböcker, språk, stil och innehåll. *Svenska Linnésällsk. Årsskr., 49 (1967):* 28-64.

GOURLIE, N., 1953. *The prince of botanists: Carl Linnaeus.* London: Witherby.

GULLANDER, B., 1970. *Linné på Öland.* Stockholm: P. A. Norstedt.

GULLANDER, B., 1971. *Linné på Gotland.* Stockholm: P. A. Norstedt.

HECKSCHER, E. F., 1942. Linnés resor, den ekonomiska bakgrunden. *Svenska Linnésällsk. Årsskr., 19:* 67-120.

HELLER, J. L., 1964. The early history of binomial nomenclature. *Huntia, 1:* 33-70.

HULTÉN, E., 1950. *Atlas of the distribution of vascular plants in N.W. Europe.* Stockholm: Generalstabens Litograf. Anslalts Förlag (2nd ed., 1971).

JOHANSSON, K., 1897. Hufvuddragen af Gotlands växttopografi och växtgeografi. *K. svenska VetenskAkad. Handl.* (N.F.), *29:* no. 1.

NATHORST, A. G., 1909. *Carl von Linné als Geolog.* Jena.

NORDSTEDT, O., 1920. Prima loca plantarum Suecicarum. *Bot. Notiser, 1920,* Bilaga.

OTTOSSON, I., 1965. Woods on the isle of Jungfrun. *Acta phytogeogr. suec., 50:* 141-143.

PETTERSSON, B., 1958. Dynamik och konstans i Gotlands flora och vegetation. *Acta phytogeogr. suec., 40:* 1-288.

PETTERSSON, B., 1965. Gotland and Öland: two limestone islands compared. *Acta phytogeogr. suec. 50:* 131-140.

PETTERSSON, B., 1966. Gammalt och nytt in Gotlands flora. *Svenska Turistfören. Årsskr., 1966:* 131-135.

SÖDERBERG, E. & KÖKERITZ, K. G., 1948. *Våra vilda Växter, 2: Öländska och Gotländska Växter.* 2nd ed. Stockholm: A. Bonnier.

SELANDER, S., 1940. En botanist på Gotland. *Svenska Turistfören. Årsskr., 1940:* 131-154.

STEARN, W. T., 1947. The nomenclature and synonymy of *Tofieldia calyculata* and *T. pusilla. J. Linn. Soc. (Bot.), 53:* 194-204.

STEARN, W. T., 1957. An introduction to the *Species Plantarum* and cognate botanical works of Carl Linnaeus. Prefixed to Vol. 1 of Ray Society facsimile of Linnaeus's *Species Plantarum,* 1753. London: Ray Society.

STEARN, W. T., 1959. The background of Linnaeus's contributions to the methods and nomenclature of systematic biology. *Syst. Zool., 8:* 4-22.

STERNER, R., 1938. Flora der Insel Öland. *Acta phytogeogr. suec., 9:* 1-170.

STERNER, R. & CURRY-LINDAHL, K., 1955. *Natur på Öland.* Stockholm: Svenkst Natur.

NOTE ON THE TRANSLATION

The first draft of this translation from Swedish into English was made by Marie Åsberg, using C.O. von Sydow's 1962 edition, *Öländska och Gotländska Resa förrättad Ar 1741,* which often differs in punctuation from Linnaeus's original of 1745, *Öländska och Gothländska Resa förrattad Ahr 1741,* and which has modernized Swedish spelling. William T. Stearn then revised her translation, using primarily Linnaeus's 1745 text and J.C.D. Schreber's German edition, *Herrn Carls von Linné Reisen durch Oeland und Gothland* (1764), and added modern binomial scientific names for the organisms mentioned by Linnaeus under vernacular names or "pre-Linnaean" mononomial to polynomial names. This addition of modern equivalent names proved a more difficult and a much more time-consuming task than had been anticipated. It could not have been done in the time available but for the modern scientific names, unfortunately without authors, provided in the notes (pp. 344-377) to von Sydow's edition and for which acknowledgement is there made to Erik Almquist, Einar Du Rietz, Rolf Santesson and Åke Holm. In his German translation, Schreber, who had studied under Linnaeus at Uppsala and become a devoted Linnaean thoroughly acquainted with the master's works, substituted Linnaean binomials for Linnaeus's original 1745 names and thus provided a convenient link, as do less conveniently Linnaeus's *Species Plantarum* (1753) and *Systema Naturae* 10th ed. (1758-59), between these and later names. Linnaeus often used in the text of the *Resa* a name slightly or very different from the binomial with a *Flora Suecica* number adopted in the index. Thus in the text of the *Resa* there occurs the name *Plantago foliis semicylindraceis integerrimis* and in the index the name *Plantago* 127 *anguina,* but in the *Species Plantarum* Linnaeus named the species *Plantago maritima,* the name it still bears. This is indicated in the following translation by an entry as follows: "*Plantago foliis semicylindraceis integerrimis* [*P.* 127 *anguina* of Index = *P. maritima* L.] covered the entire field". For plants common to both the British Isles and Scandinavia, the nomenclature adopted mostly follows J. E. Dandy's *List of British vascular plants* (1958) and M. Parke & P. S. Dixon, "Check-list of British marine Algae" in *J. mar. biol. Ass. U.K., 48:* 783-832 (1968), but use has also been made of other works, notably J. Lid's *Norsk og Svensk flora* (1963) and those by Nils Hylander. For birds, butterflies and fishes the nomenclature follows R. Peterson, G. Mountford & P. A. D. Hollom, *A field guide to the birds of Britain and Europe,* 2nd ed. (1965), I. G. Higgins & N. D. Riley, *A field guide to the butterflies of Britain and Europe* (1970), A. Wheeler, *The fishes of Britain and North-west Europe* (1969) and T. Gislén & H. Kauri, "Zoogeography of the Swedish amphibians and reptiles" in *Acta vertebratica, 1*(3) (1959).

The names of places have been modernized as in von Sydow's edition. Thus the "Öfwerste-Qwarn" of Linnaeus's time is now spelled "Överstevarn"; "Calmar" is now "Kalmar"; "Afwe" is now "Ave"; "Hwittlanda" is now "Vitlanda"; and so on.

The runic inscriptions copied by Linnaeus on Öland and in Småland have been reproduced direct from his work but are followed by a transliteration of his version (cited as "Linnaeus") into Roman characters, a transliteration of the same inscription when extant as rendered in *Sveriges Runinskrifter,* Vol. *1 (Ölands Runinskrifter)* or Vol. *4 (Smålands Runinskrifter)* and a translation into English, for all of which we are much indebted to Dr Helmer Gustavson of Riksantikvarieämbetet och Statens Historiska Museum, Sweden.

It should be noted that, when Linnaeus wrote, Sweden, like England, had not yet adopted the Gregorian reform of the calendar. Hence *the dates of Linnaeus's journal are Old Style, i.e. 11 days behind the now accepted New Style.* For example, the date "Jun. 1" on which he crossed from Kalmar to Öland corresponds to the modern 12th of June.

The two anthologies by Bertil Gullander, *Linné pa Öland* (1970) and *Linné pa Gotland* (1971), did not become available until after the present translation had been made but are of special interest for their extracts from Linnaeus's unpublished manuscript belonging to the Linnean Society of London, their marginal notes and illustrations and their photographs of Öland and Gotland.

W. T. S.

The Journey to Öland and Gotland

CARL LINNAEUS

1741

PREFACE

In the Riksdag [the Swedish Parliament] of 1741 the Estates were pleased to decree that I should make a journey to several districts within the Kingdom, especially to Gotland, Öland, Västergötland, Kinnekulle, Halleberg and Hunneberg, Mösseberg and Olleberg, Billingen and other places. Instructions consisting of the following items were issued by the office of Manufactory Administration on 27 April at the request of the State Commission for Trade and Manufacture:

(a) To search for plants and grasses that are suitable for making dyes, and to instruct the farmers in their proper use.
(b) To investigate the occurrence of clay or earth that might be used for pottery, tobacco pipes, fulling, etc.
(c) To investigate which plants, hitherto imported for pharmaceutical purposes, could be found within the Kingdom.
(d) To inform myself of things pertaining to the *Historia Naturalis Patriae* [natural history of our country], such as trees and plants, animals, birds, reptiles, etc.
(e) To keep a careful and accurate diary and submit it to the authorities on my return.

On the 15th of May the same year the journey began. I passed Södermanland, Östergötland and Småland. From Kalmar I sailed to Öland and followed its west coast southwards to Ottenby, then northwards up the east coast, paying a visit to Blåkulla (or Jungfrun) and afterwards sailing from Öland to Gotland. There I followed the west coast from Visby to Fårö, then the east coast to Hoburgen in the extreme south, returning along the west coast to Bursvik, Ekstadt, both Karlsö islands, Klintehamn, Roma and Visby. Finally, on leaving Gotland, I returned to Stockholm via Öland, Kalmar, Växjö, Visingsö and Örebro, finishing my travels on the 28th of August.

I will briefly summarize what I accomplished during my journey, and the extent to which I was able to follow my instructions:

(a) The dye-yielding plants discovered during the journey have been sent to the Royal Academy of Science, as can be read in the *Handlingar,*

1742: 20. ["Förtekning at de förgegras, som brukas på Gotland ock Öland", *K. svenska VetenskAkad. Handl., 1742:* 20-28.]
(b) As regards clay, very little has been achieved since in Gotland and Öland it is mostly mixed with lime and chalk. The clay that is found between the sandstone layers in the quarries at Bursvik and Grottlingebo, however, is of a special kind. It is dry, somewhat slatey, dense and fine and disintegrates into cubes. It removes stains and possibly could be used for fulling.
(c) Several pharmaceutical herbs, hitherto imported and not known to grow wild in Sweden, have been found, such as *Scordium, Eryngium, Stoechas citrina, Ruta muraria, Ebulus, Bellis minor, Victorialis, Cichorium* and *Lithospermum* amongst others.
(d) As regards Natural History, I have found more than could reasonably be expected during this journey. In the *Handlingar* of the Royal Academy of Science 1741: 179 [i.e. pp. 179-210], I have enumerated one hundred plants, discovered on this journey but unknown in Sweden before. As regards ore, there is little of it in Öland and Gotland, although there are a great many corals and petrifactions. A considerable number of animals, fishes, amphibians, reptiles and especially insects and birds has been collected.
(e) I now have the honour to submit my diary to the public, so that it can be seen with what enthusiasm I have carried out my instructions.

I have noted mileages and lodging-places briefly. I have deliberately avoided harsh judgements and critique of mistakes I have seen now and then, in order to be of use without hurting anyone. Things have been presented briefly without many comments or reflections—which are self-evident, when the data are correct—for the sake of brevity, the most pleasant manner of writing. I have purposely described much that is common in Sweden because it is rare in other countries.

Antiquities, such as castles, burial mounds, barrows and runic stones, have been noted briefly. I have recorded a great many runic inscriptions in Öland in order to show others where they can be found rather than to give any reliable copy of them. On one hand, this is not my field, on the other, I did not wish to be delayed by the long process of removing the difficult moss which covers all the runic stones of Öland. My reader should thus regard these copies as a hasty piece of work and short annotations. As regards the runic stones in Gotland, I have only mentioned those I have seen without copying one letter, since I have been informed that the learned Bishop G. Wallin has copied them all very meticulously and is about to publish them.

Economic matters have been described in more detail. The habits of the farmers, their clothes, buildings, their farming methods, fields, meadows, forestry, tar boiling, charcoal burning and quarrying have been described, and I have now and then briefly indicated how methods could be improved.

As for mineralogy, I have reported rather extensively on limestone, sandstone, drip-stone, shifting sands and other things not previously described.

In botany, I have made many observations in various places, which I hope will be of benefit to our fatherland. I have found the weed that is burned for making soda; the sedge which is used in Gotland for making long-lasting roofs I have described—it ought to be easy to introduce it into our vast marshlands; the marram grass, which retains the shifting sands in a remarkable way; darnel, which makes people almost blind when put into a drink; plants that give milk and butter a garlic taste; ramsons, that keeps weeds out of hop-gardens; hawthorn, that should be planted were no other tree can endure; the *"Tok"* [*Potentilla fruticosa* L.], excellent for low garden hedges; charlock, describing its strange nature and the harm it does; Swedish Hay Seed, so excellent for pastures that I have never seen its like and the Öland Hazel, used for low hedges. I have recorded how the oak furnishes a chronicle of cold and mild winters for the last 300 years, and how oaks are treated to make them grow straight trunks; how the seaweed is used as manure on the fields; which plants thrive on the most dry and meagre soil, and which plants indicate that the soil is marshy; where oil can be made and from which plants; where the heather does not thrive, and why; which plants are harmful to the sheep, and the delightful herbage plant that is the best food of all for sheep; the herbage plant that cattle feed on at the seashore (*Handlingar* of the Royal Academy of Science 1742: 191).

It is shown what kind of plants Agh, Alfwarloek, S. Brittae loek, Rams, Keipe, Kräkel, Madra, Mannablod, Selting, Tarald, Took, etc. are.

Within zoology, I have described how seals are caught, how deer are kept out, how rooks are caught, how the eiderduck is mismanaged and how mallards are trapped; how cod and flounder are fished; how remarkable are the razorbill, the avocet, the golden plover, the turnstone and the oystercatcher as well as numerous insects, the use of which is still largely unknown to us, although the knowledge of them is delightful for those who have read what Réaumur, Swammerdam and Frisch have discovered for the world.

How my journey has served to increase the knowledge of *Historia Naturalis Patriae* will be evident from the books I am editing at present—these will surely show that the *Historia Naturalis Patriae* regarding plants, animals, birds, fishes, worms and reptiles has been recorded more carefully in Sweden than anywhere else in the world.

I have made several new observations about the physical properties of streams, lakes, salt and bitter wells, springs, lime water, mountains, caves, tussocks, etc.

Local remedies for several diseases such as swellings, wounds, lung-inflammation, hysteria, whitlows, dysentery, spleen, etc. have been noted.

Superstitions have also been recorded now and then, more to entertain my reader than for any other reason.

A great many plants and insects have been described in Latin, partly for the sake of brevity, partly because the *termini artis* [terms of art, technical expressions] are not established in the Swedish language, but above all to enable foreigners to peruse that part of the work, thereby encouraging a desire to acquaint themselves with the rest of it, especially as no Swede will probably take the trouble to read these descriptions if he is not already familiar with the study of such.

To teach the inhabitants and the farmers to recognize and prepare dye plants and pharmaceutical plants was found to be very difficult. The short summer necessitated a hurried journey, in order that I might have the opportunity of visiting many places and making as many discoveries as possible. I will exert myself to teach the students at the university all such things, so that when they become clergymen in the countryside, they may instruct their parishioners to best advantage.

The maps of Öland and Gotland are made by the learned inspector Jacob Faggot of the Land Surveying Bureau. He has given them to me in order to clarify this little book.

I shall not forget to mention six young gentlemen, who at their own expense accompanied me on the journey, learning to make observations and training themselves and busily helping me with my investigations. The were

P. Adlerhielm, Notary, *Kgl. Bergskollegium*
J. Moraeus, Auditor, *Kgl. Bergskollegium*
H. J. Gahn, Iron master
G. Dubois, Student of Medicine
Fr. Ziervogel, Royal Pharmacist
S. Wendt, Student of Medicine and Botany

My style of writing is very simple, which will doubtless earn me a low reputation among the nightingales of Pliny*. Language adorns a science as clothes adorn the body. He who cannot do honour to his clothes must needs let them do honour to him. If elsewhere in the world only eloquent doctors had been allowed to write, less would be known than there is today. Should my countrymen be pleased, something more delightful will, with God's help, be forthcoming; if not, my labours shall not be the source of further criticism.

* This metaphorical allusion to masters of eloquence is based on the wonderful description of the nightingale and its song in Pliny's Natural History (book 16, chapter 29), which refers to "a nightingale which settled upon the mouth of *Stesichorus* the Poet and there sung full sweetly; who afterwards proved to be one of the most rare and admirable musitians that ever was" (Philemon Holland's translation, 1601)—W. T. S.

Biol. J. Linn. Soc., 5 (1973

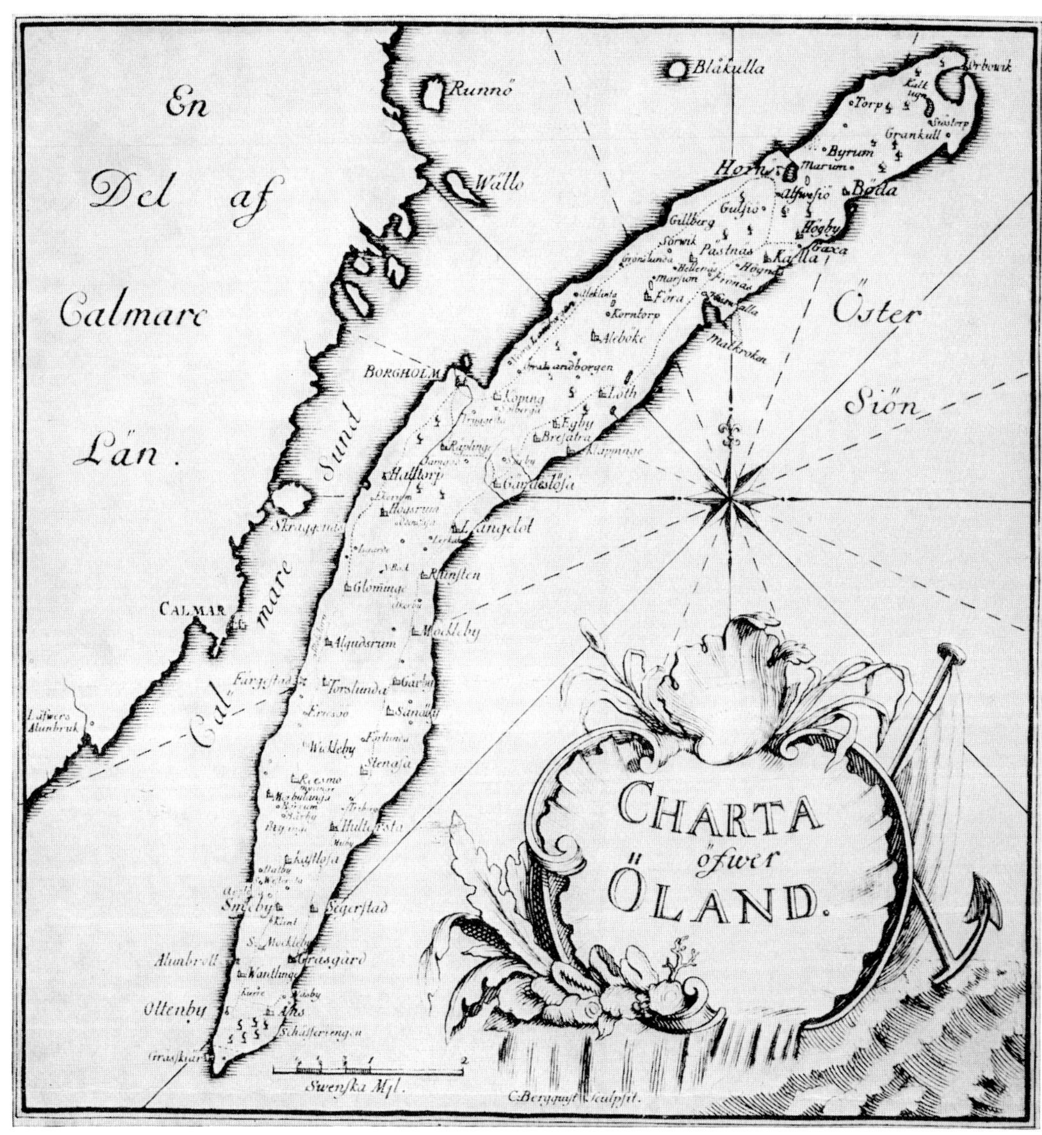

Map of Öland by Jacob Faggot
(reproduced from Linnaeus's *Öländska och Gothländska Resa,* 1745).

(*Facing p.* 23

The Journey to Öland

CARL LINNAEUS

made in 1741

1741, May 15 [p. 1]*

We left Stockholm in the most pleasant period of spring, at 11 a.m. on the 15th of May. The sun was shining, but it was quite cold.

Spring should be measured according to climate and temperature rather than by the calendar; it had advanced so far that the maples were in flower but had no leaves, the first birch leaves had appeared, and there were little buds, like wild strawberries, on the smallest twigs of the fir trees, but there was no pollen yet. Linden, oak and aspen still rested in their winter sleep.

There were no flowers except *Draba prima* [*Erophila verna* (L.) Chevall.], *Caltha* [*C. palustris* L.], *Glechoma* [*G. hederacea* L.], *Leontodon* [*Taraxacum*], *Hepatica* [*Anemone hepatica* L.], *Anemone secunda* [*Anemone nemorosa* L.], *Oxalis* [*O. acetosella* L.], *Adoxa* [*A. moschatellina* L.] and *Salices*.

Among the birds, the cuckoo had begun to call and today we saw the first barn-swallow.

SÖDERMANLAND

Our journey proceeded through Södermanland, but the landscape was still the Uppland type with low, blue hills and patches of forest and beautiful cornfields between them, although the crops were somewhat uneven.

Outside the town we met a company of soldiers with brand-new clothes and a healthy and robust appearance.

After 1½ miles' journey we arrived at Fettja, where we had to wait a long time for horses, and after another 2 miles we reached Södertälje.

Södertälje is a little town in a valley between two hills in a sandy area which stretches out along the valley. On one side is the lake Mälaren, on the other a bay of the Baltic, which silts up every year so that ships cannot come closer to the town than a quarter of a mile.

From here the landscape changed and looked more habitable with larger valleys, cornfields and meadows; the sea could be seen from the road in several places.

---

* The page numbers following the dates refer to the page of the 1745 *Öländska och Gothländska Resa* on which the entry occurs.

We reached Pilkrog in the evening after 1½ miles' journey and hurried along another 2 miles to Åby; since we were not yet accustomed to travelling, we spent the night there, from midnight.

May 16 [p. 2]

It was a warm and pleasant day after a cold night with intense frost.

Wormwood grew abundantly around the farms here.

There were more dog roses along the road from Stockholm than anywhere else in Sweden; they had a lot of grubs on them, the *Bedegvar* of the apothecaries; some of them had not yet been transformed to *Ichneumones.*

Bryum capsula nutante turbinato ovato, calyptra reflexa quadrangulari [*Funaria hygrometrica* Hedwig], which changes appearance in warmth and drought, made red patches here and there on the ground.

We left the main road and rode in the direction of Trosa, and on both sides of the road we saw a reddish hue on the shattered rock as an indication that the bedrock contains iron.

Gäddeholm, a beautiful and well-built manor house, was on our way; we visited its garden, where many flowers were out, such as *Tulipa, Crocus, Narcissus luteus, Hyacinthus botryoides* [*Muscari botryoides* (L.) Miller], *Fumaria bulbosa cava & Solida radice* [*Corydalis bulbosa* (L.) DC. & *C. solida* (L.) Sw.], *Primula, Bellis* [*B. perennis* L.] and *Corona imperialis* [*Fritillaria imperialis* L.] ; there were also lilac and privet hedges, maples and box-trees. We also saw nicely trimmed *Taxi* or yew trees. Fig trees and *Thuja* were the rarest plants. There were avenues of willow trees in flower. They were all male, thus they could not spread their seeds all over the place and make it untidy with their woolly seed-vessels during the summer.

Trosa town is situated at the inner end of a bay, and small ships could sail up to the market-place and the bridge; this is the only remarkable feature about this town, which is quite small. Its situation, a quarter of a mile away from the main road, has probably been an obstacle to further development.

We had the opportunity to describe a little butterfly which has not been seen abroad: *Papilio hexapus alis divaricatis denticulatis nigris albo punctatis* [*Hesperia malvae* (L.)] . It was very small and six-footed and held its wings in a strange position, like that called "Ekeblad" [oakleaf] ; the wings were soot-coloured with white spots on both sides, paler on the under sides and edged with teeth formed of alternating black and white bristles.

After a 2 mile journey from Åby on a hilly road we reached Svalbro, where we had to remain in the cold during the whole night for lack of horses, beds and night-lodging.

May 17 [p. 4]

At 8 o'clock in the morning, as soon as our horses were ready, we continued our journey, leaving the beautiful Svalhammar on our left hand, and after 2¼ miles we arrived at Nyköping in heavy rain.

Nyköping is a beautiful town with neat houses, broad streets, fresh air and several stone buildings. We saw the dilapidated castle and its ruined walls, and the prisons of the princes Erik and Waldemar; this place is at present the residence of the local government authorities and the city council, and there are also a few shops.

Among the walls of burnt-down houses in the town, there was one made of Dutch clinker, which was in a good condition in spite of its having stood more than 20 years in the open air since the last war. The clinker does not absorb as much water as our bricks and the bricks are smaller and better burnt, and white inside in contrast to our brick, which is red and contains iron.

A skilfully constructed greystone bridge passes over the river which flows through the centre of the town.

There are many good waterfalls within the town, which drive flour mills, paper mills, rolling mills and a brass foundry.

When the sun shone on the mist of water-droplets which rose from the waterfalls, one saw beautiful rainbows.

The flour mills had Rhineland millstones, and one single mill had ten pairs of them. In one of the mills there was a picture of Saint Anna, which the people considered it dangerous to touch, and they also told us about the disasters that had been caused by irritating or moving the saint. The flour from Nyköping is said to be the best in the country, which to some extent is obviously due to these good mills and millstones, but the most important factor is a good grain; the best wheat grows in Södermanland.

We visited the saffian* factory, where we saw how the hides are prepared and tanned to win the credit of a whole nation; Christ. Ickorn, who showed it to us, has constructed it himself without public support; he was however worried about the paper mill since its existence prevented him from obtaining a stamper. Sawdust from the sawmill was employed as an absorbent to dry the hides and make them easier to handle.

We also visited the brass foundry, where we saw the casting of brass and the thread drawing; I do not understand why it should be necessary to import the stones for casting, since they consist of mica, quartz and spar and could equally well be found within the country. *Nihilum album* [zinc oxide] was not collected although present in great quantity; the only thing needed is an old iron cauldron. It could be sold to the apothecaries and also exported at a price of a *daler* a pound or even less. The price in the pharmacies is one *styver* for a *lod.*

The starch factory, which has been built by the surgeon, Ziwert, was also visited; a substance was shown to us which was very similar to Prussian blue and made without safflower [*Carthamus tinctorius* L.].

The paper mill with its rags and bleaching troughs, glue water and roller, pressers and stamps was visited in the afternoon.

Ziwert told us that the antiquities, the barrows and stones described by Sundler in his *Diss. de Nycopia pag. 150* and cited by Tuneld in *Geograph. 62* are quite worthless.

Marsh rosemary (*Ledum*) [*L. palustre* L.] was mentioned as an excellent remedy against lice. "Stäckra" [*Oenanthe aquatica* (L.) Poiret] grows abundantly in Södermanland and is also said to do more harm to the horses here than in other places. Dean Pihl told us that it is not dangerous when it is fresh, only when it is dry, but this may be so because the horses recognize and avoid it by smell and taste when it is green.

---

* Saffian was goat–or sheep–skin tanned with dried leaves of Sumach (*Rhus*) and dyed bright red or yellow.–M. A.

May 18 [p. 6]

We left delightful Nyköping on a good road between cornfields. It was warm in the sun and the larks warbled in the sky. The winter rye was somewhat thin, better on the hills but all spoilt in the valleys. We saw a great many barrows on our left hand shortly before we arrived at Baerchner's place called Berga.

Snakes are not sufficiently well described, either in Sweden or abroad; therefore we intend to describe those we found in order to enrich the *Historia naturalis patriae.* Here we found a *Coluber* [*Natrix natrix* (L.), grass snake] with a grey body and ovate scales, smaller on the back, with an elevated line on each scale; these were all uniformly coloured except the tips, which were black, giving it a mottled appearance. The underside was covered with scales which were as broad as the belly or half of the body; these scales were black with lighter sides and still lighter at the head end. The head was black and had several sutures. The jaws had black transverse stripes. The nostrils were small. The area around the ears was nodular and protruding and had a large white spot. There were two rows of equally sized teeth. The tongue looked like two black threads. This snake which belongs to the grass-snake genus does not give any dangerous bites.

After a 1½ mile journey we changed horses at Jäder's inn. A snow plough, a clod-crusher and a roller lay here. The clod-crusher was made of two heavy parallel boards on edge, securely joined with cross bars; it is used to flatten the field and crush the clods that remain after the rolling. I intend to describe even the most simple and common agricultural and household devices, not so much for my countrymen as for the foreigners. Such information may be of use, however, in the *Economia privata.*

After travelling another 1¼ miles to Vreta on a good road we noticed a change in the soil.

Kolmården is a vast forest with spruces, fir trees, rocks and boulders, hills, mountains and precipices of which Skutskärsbrinken is one of the steepest. Everywhere along the road there were rocks of a scarlet colour, which when rubbed changed to intense yellow and smelled like violets. Therefore this stone is called violet stone, although it is certainly not the stone that smells, but a fine moss (*Byssus*) which colours it. This *Byssus* does not consist of threads like the foreign one, but of a fine powder.

After 1¼ miles we arrived at Krokek's inn and saw the church which is built on the ruins of Svintuna monastery.

Mountain currants [*Ribes alpinum* L.] grew abundantly here; they are similar to red currants but have erect flower clusters, broader bracts and narrower leaves, edged with short bristles which end in a gland.

Here at Krokek we saw a little butterfly [? *Callophrys rubi* (L.), the Green Hair Streak] similar to *Argo* but without spots, with intensely blue wings.

Cochlea testa utrinque convexa subtus perforata, spira acuta, apertura ovata transversali [*Campylaea lapicida* (L.)] was found among the ruins. It is uncommon in Sweden and rare abroad and easily distinguished from other species since the shell has sharp margins, not rounded.

Cerambyx nebulosus, antennis corpore longioribus, thoracis punctis quaternis luteis [*Acanthocinus aedilis* (L.)], an insect which Petiver received from Russia and described as a rarity, was found here in Kolmården. It was iron grey, and one inch long; the wing cases were obtuse, covered with grey hair and dotted with small tiny spots, and there was a darkish transverse furrow across them. The breast, which was of similar appearance, had a tooth on each side and four yellow spots. The head was also iron grey. The eyes looked like shagreen and there was an eye-like spot behind the antennae. The antennae were five times as long as the body and made of eight joints, each of which was grey below but black and thicker at the outer end; they grew gradually shorter closer to the head. Abdomen and feet were grey. The feet had four joints, the first two crescent-shaped, the third cloven, as in a goat's foot, black with white bristles, and the fourth oblong with sharp claws. The wings were dark but transparent and had soot-coloured veins. In relation to its size, the creature was quite strong.

Here at Krokek inn is the border between Södermanland and Östergötland.

ÖSTERGÖTLAND

From Krokek to Åby we had a 1½ mile journey through a like landscape and hills, of which Glasjöbacken above and Rotbrinken below were the most remarkable. Soon before we arrived at Åby inn, we left Kolmården. It is remarkable that, although Kolmården is all rocks and mountains, its edges are sandy without any stones and, further down, the soil is all clay. Could this be the work of the sea? When we left Kolmården the landscape changed and we saw Östergötland with its vast cornfields, villages, roads, the Baltic and Norrköping.

We arrived at Norrköping in the evening after a ¾ mile journey from Åby.

May 19 [p. 9]

We spent all this day in Norrköping in order to look at the town and its factories, and also to avoid the soldiers who were marching in, since their arrival would surely have made it impossible for us to get horses. The sugar factory built by Alderman Lindstedt and his partners was the first to be visited; the house had eight storeys including the attic. Here the coarse and unrefined raw sugar was pulverized and boiled in water, diluted with lime-water, mixed with ox blood or egg white, skimmed and poured into inverted cone-shaped moulds, perforated at the top; from these a syrup trickled down into a bottle; this was repeated, and then the mould was covered with a white, dough-like French clay like a lid. It is strange that there should be no such clay in Sweden, but it has to be imported. The only function of the clay seems to be to protect the sugar from too rapid desiccation. If the clay is removed and only the tip of the sugar-loaf is still dry, it does not turn white. The scum which is skimmed from the boiling kettles and the syrup that flows from the moulds is used for making treacle. The lime-water which contains lime salts is as harmful to a man's body as it is good for refining sugar.

Then we visited the brass foundry, but since no work was going on, we left immediately.

The cloth factory was shown to us by Commissioner Urlander; we saw how the wool was spun, woven to cloth and carded with teasels (*Dipsacus*). It delighted us greatly that Swedish men can wear Swedish-produced clothes, and also that many poor people can maintain themselves through their own work. But it vexed us that teasels should be imported when they are so easily grown in our gardens, where they sow themselves once planted. We know well that the cultivated teasels grow softer and thus less suitable for this use than the wild ones, but this could be avoided if they were planted in a stiff but rich soil in an open place exposed to the winds. Sawwort (*Serratula*) [*Serratula tinctoria* L.] was used for dyeing blue. It costs 5-6 *daler* a *lispund,* although it grows wild around Uppsala and Stockholm and in Östergötland, Öland and Skåne. For yellow, weld (*Luteola*) [*Reseda luteola* L.] is used, which is also imported although it grows like a weed everywhere in Lund–it should be cultivated where it grows so readily. For black, *Vitriol* and bearberry twigs are used, instead of the foreign sumach or *Rhus.*

In the afternoon we visited a tobacco factory where we saw many little children earning their living; some of them made plaits, twisting strips of tobacco soaked in a tobacco liquor, the composition of which was an *arcanum* here; others spread out the tobacco leaves, still others dressed the plaits with leaves of a few quarters' length. Some removed the midribs from the leaves, others were twisting tobacco into ropes, still others were cutting it. It surprised us to see the force of a worm screw in the tobacco press.

The river which passes through the town meanders inside as well as outside it, like a horse-shoe. There were a few ships on it, and four bridges over it. Salmon was fished within the town boundaries. In the river there were tench, lampreys, ides [a kind of carp] and sometimes sturgeon. The fishing was done in a nice and easy way by two men in a little boat. One was rowing, the other stood at the prow with a bag net with a flat bottom and there were often many fish in it when it was pulled up. In some boats there was only one man with a net and nobody at the oars, and the boat was left to drift with the current, until the man left his net and returned to the oars to row upstream and start over again. In the river near the brass foundry there were some islets in the middle of the torrent which were impossible to reach. Willows grew on them and there were nests of birds like magpies or crows in the trees. The birds were said to look like crows but smaller, with red bills and red feet. They sounded almost like crows. They were away all day feeding but returned to the nests in the evening. The inhabitants had no name for them, since they arrived here only four or five years ago and had increased in number since then. With very skilful planning a safer shelter could not be found in the middle of a town than that chosen by these birds, especially as they are migratory and only pass the summer here. Mr Stockenström promised to shoot one and send it to the Royal Academy of Science (he kept his promise and the bird turned out to be a rook [*Corvus frugilegus* L.]). (See also June, 3 and 5.)

The gardens were full of tobacco plants, *Fritillaria, Auricula* and *Portulaca,* all in flower, of which the last-named had a four-cleft style.

The town was big, with beautiful houses, broad streets, three churches, four squares, many gardens and two mayors. Nightfall prevented our seeing more factories and other remarkable things in this large and delightful town.

May 20 [p. 12]

In the morning we left Norrköping in cloudy weather, but at 8 o'clock the sun was shining and the day became quite hot, and we suffered much from the road dust.

The roads are beautiful here in Östergötland planted both sides with willow trees [*Salix fragilis* L.], which grow best in a wet soil; where the soil is dry, sallow, which is a *Salix* too and similar in appearance, should be planted instead. We noticed the whole day that there were only male willows and no females, so that there are no woolly seed vessels to litter the roads. If there were plenty of bees here, they would have a delightful task with these flowers in the early spring. These willows were very brittle at the joints, and when the twigs were bent they cracked at the joints but were tough and pale everywhere else. The leaves were lanceolate, pointed, serrated, with a smooth shiny upper side and fine hairs on the underside. The flowers had two stamens, and anthers with a yellow honey gland, and the sepals were slightly downy.

In a few places another kind of *Salix,* called "Jolster" (*Salix floribus tetrandris*) [*Salix pentandra* L.], grew among the willow trees; it was smaller, the leaves were egg-shaped, glossier, more finely serrated and not downy. The branches were red and not at all brittle, and there were both males and females. On one of these was a recent *Rosa Salicina* [willow gall], and it could be clearly seen from it that Ray was wrong when he said that these roses are produced by worms; it was obvious that it was nothing but a *plenitudo* or a double flower, where the stamens had been transformed and the calyces of the male flowers had attained a monstrous size.

On the road we met a company of soldiers from Jönköping on their way to Stockholm and the galleys.

We passed Löfsta, the estate of the de la Gardie family, situated on a hill top on our left and Gransjön on our right hand.

The spruces were in flower and as soon as they were touched a cloud of dust rose from them; their endless amounts of male flowers, which looked like flesh-coloured strawberries not long ago, were pale now, having dehisced; there were a few drooping scarlet female flowers with scales bent backwards and hollowed like spoons to receive the pollen and direct it to the red lanceolate styles.

On the spruces, were *Cimex* grubs, the size of a bedbug, but stouter and rounder, dark altogether, with black eyes and four longitudinal rows of black points.

Kimstad church was passed on our left hand; two riders cut out of metal sheets were riding towards each other on the roof ridge.

Here, as in other places, we met large herds of oxen on the road, on their way to the slaughter houses in Stockholm.

Around Norrköping there were large cornfields, small junipers, no forest, a few deciduous trees, and little stone.

There were quite a lot of oaks here, but they were mutilated and ill-treated; the thick branches had been cut off at a hand's breadth from the trunk, but too late; small twigs sprouted around these cut-off branches, the crown of the tree was too small and insufficient for such a big tree, and many oaks were withered and had died.

The flowers were: in the ditches *Hottonia* [*H. palustris* L.] and *Ranunculus capillaceo folio* [*R. aquatilis* L.]; in the marshes *Caltha* [*C. palustris* L.]; in the meadows *Primula lutea* [*P. veris* L.]; in the shrubbery *Anemone nemorosa* and in the tussocks *Juncus capitulis psyllii* [*Luzula campestris* (L.) DC.].

After 1¼ miles we came to Brink inn.

At Roxen the road crossed the lake, and along the road on a little island was

Norsholm's farm, neat and well built. In the fields were yellow little birds (*Motacillae flavae*) [yellow wagtails] and in the wood *Montifringilla calcaribus alaudae* [*Fringilla montifringilla* L., brambling].

After 1¼ miles we came to Kumla inn.

The limestone in the quarry of Kumla was all white, dense and coarse, but there was also a flint-like stone with very fine grains. It was quarried after blasting the rock, which had several fissures, and these fissures were often joined with beams, elevated some three fingers' breadths from the rock. On the ground we saw some limestones glazed to a greenish blue glass by the fire.

Saxifraga tridactylites tectorum [*S. tridactylites* L.] was in full flower. We noticed that pairs of stamens were bent together, touching one another's anthers above the style, so as to rub off the pollen; later they separated but two other stamens took their place. This is an incomparable experiment in the nuptials of plants, which has not been observed before. The whole plant was covered with short white bristles, and a red gland at their top excreted a clammy fluid which made the plant sticky.

Potentilla quinta [*P. verna* L.] was in full flower on the hill and we saw *Polytrichum calyptra striata sursumque pilosa* [*Orthotrichum striatum* Hedwig] with its *capsulis sessilibus, subglobosis, rectisque calyptra hispida, arcta, acuminata, totam capsulam involvente.*

Festuca panicula secunda coarctata aristata, culmo quadrangulari nudiusculo, foliis setaceis [*F.* 95 *ovina* of Index = *F. ovina* L.], a grass which is common on our barren hills, and which has been an enigma to most botanists, flourished here; the roots were hair-like, blackish, long and perennial; the leaves were gathered together at the root like a broom; they were one inch long, thick as bristles, slightly compressed and rounded; they were rough when stroken downwards with the finger but smooth upwards. The stalk was a quarter-ell [15 cm] tall. The panicle was narrow and compressed, reddish and nodding. The spikelets were lanceolate with glumes without awns, tapering towards the tip which carried four flowers with awns.

We had the opportunity here to describe a hawk [*Accipiter nisus* (L.), sparrow-hawk]; it was similar to a cuckoo but slightly bigger, had long, narrow feet, black claws and a small bill; the crown, back, tail and wings were iron grey; the underside was all white with grey wavy lines across and each feather or breast had a hardly visible brown spot at the tip; the tail feathers were iron grey with six dark cross-lines; the under tail-coverts were snow white.

Mordella fusca opaca, elytris punctatis [*Cimex clavicornis* L.], a little insect, which devours small plants in spring, was plentiful here. The body was black, the wing cases had hardly visible spots; legs, feet and the inner part of the antennae were yellow brown; the antennae had ten joints; the hindlegs with which it jumps like a flea are very strong, and it is of the same size.

Everywhere in the villages starlings had built their artful nests with double entrances. Strange bird that avoids the woods and exposes itself to continuous danger living among people!

One of the king's hunting grounds was seen along the road at a distance of 1½ miles from Kumla and ¼ mile before we arrived at Linköping.

Linköping is no big town, the houses are small, the streets uneven and there is nothing remarkable about it. The cathedral is one of the biggest in Sweden. Here were the residences of the bishop and the district governor, a high school and a school. On the market-place were some large stone slabs, at the place, according to report, where King Charles IX had his Council executed.

The district governor told us that a strange tree grows on the skerries here, which, according to his description, I reckon to be a *Taxus* or yew.

We did not stay long in Linköping. Outside the town we passed through Malmen, a forest which is very similar to the royal hunting grounds at the other side of the town.

Peas were sown on some fields here.

Ranunculus foliis submersis capillaceis [*R. aquatilis* L.] and *Oxalis* [*O. acetosella* L.] closed their flowers towards the evening.

After a journey of 1 mile from Linköping we reached Bankeberg. Evening came, and we hurried on.

After a further 1¾ miles we arrived at Mölby in a red sunset. Mölby inn is remarkably well situated where the roads from Ekesjö, Skänninge, Vadstena and Linköping cross. If the innkeeper were as good as the situation, at a place where so many people pass and where so many meals are eaten, the inn would not be what it is. We neither could nor wanted to spend the night there, but left the place and arrived at Dala at midnight after a further ¾ mile. At 4 o'clock in the morning we came to Hester after a further 1¼ miles, and tired after our night-wake we rested here three hours.

May 21 [p. 17]

A quarter of a mile from Hester we saw the huge boundary stone between Östergötland and Småland on our right.

SMÅLAND

Before long we arrived at a gorge with steep sides, covered with rocks and wood. This gorge is the border between Östergötland and Småland, just as Kolmården is the border between Södermanland and Östergötland. The common bedrock is a red feldspar.

The flowers in this district were *Ranunculus bulbosus, Viola tricolor, Chrysosplenium* [*Ch. alternifolium* L.] and *Cerastium* [*C. semidecandrum* L.].

Chrysosplenium [*Ch. alternifolium* L.] was common in the marshes and had alternate leaves. *Cerastium floribus pentandris* [*C. semidecandrum* L.] had petals which were unevenly distributed in three parts. There were mostly five stamens, sometimes more. Five styles. The entire plant was downy, and every little hair ended in a little gland which secreted a sticky fluid.

Grey alder (*Alnus folio incano* C.B.) [*A. incana* (L.) Moench] grew abundantly on the mountains; it is the same species as our common alder, but the altitude changes it and gives it a white bark and leaves which are acute at the tip and not glossy. The leaves of the grey alder are collected in the summer as winter-fodder for sheep, which are said to prefer this leaf to common alder.

We saw farm girls collecting morels [fruit-bodies of morchella] in the woods; they were big and beautiful and one pottle of morels was sold for four *styver.*

Lake Sommen was on the left-hand side of the way.

After two miles of difficult riding on very hilly roads we arrived at Säthälla.

The farmers had a rather pronounced accent here.

The knawel [*Scleranthus annuus* L.] was in flower and its linear filaments were bent over the styles.

The junipers, all female, were also in flower with three styles and three almost overlapping scales.

Menyanthes [*M. trifoliata* L.] was in flower in the marshes. Its calyces were nicely downy, like an American *Hypericum* called *Lasianthus*.

Papilio major caudatus ex nigro & luteo variegatus Petiv.mus.35 n.328 [*P. machaon* L., the swallowtail], the biggest and most beautiful of all Swedish butterflies was found in the gorge.

Horses, which lay dead along the road, showed us by their teeth the distinguishing character of their species: there were six front teeth, six molars on each side and one eye tooth, which was separated from front teeth and molars by a small empty space.

Equisetum [*E. arvense* L.] was in flower in its own way, and its fine pollen danced for us when put on a piece of paper.

The road was for us very little better, though more and more leafy with birches.

After 1¾ miles we arrived at Berga. From here on the road was slightly better.

Burned-out patches of woodland (Svedjor, which Smålanders call fällor or lyckor) now began to be seen on both sides of the road, mostly green with an excellent rye. We saw some of the burn-beating today, since the best time to burn is after a long drought when rain is expected, so that the wood will burn well and the ashes be retained by the rain. It is remarkable to see how these fires can chase away clouds and rain—every child can tell that when the sky eventually becomes over-cast after a long dry period and if the farmers start to burn their lands, then the sky clears up, the clouds disappear, and torrents of rain fall in the next parish, while the burners receive nothing.

There was a runic stone at Söderängsbacka near the road 1½ quarter miles from Berga on the left hand. We read the following letters on it

ᚴᛚᛅᚴᛦ·ᚱᛁᛋᚦᛁ·ᛋᛏᛁᚾ·ᚦᛅᛁᛋᛁ·ᛅᚠᛏᛦᛋ····

ᛋᛁᚾ·ᛚᚢᚱ·ᛒᛁᛅᚢᚱᚾ··········

(towards the road) (on the south side)
Linnaeus: *klakR . risþi . stin . þaisi . aftR . s . . . sin nur . biaurn*
Sm 133: *klakR : risþi : stin : þaisi : aftR : sun : sin : bur : biaurn*

"Glöggr (or Klakkr) raised this stone in memory of Torbjörn, his son."

After 1¾ miles' journey we arrived at Ekesjö at 8 p.m.

May 22 [p.19]

Ekesjö town is not very large but densely built without any splendour.

We went to church, because it was Intercession Day today, and afterwards we left Ekesjö at 1 p.m. The weather was quite warm and small white clouds floated here and there in the sky, but the swallows were hunting insects high up, telling us that we would have no rain.

The soil, which yesterday was becoming covered with heather, was likewise so today and also sandier. The birch trees were all green now. The aspens had recently put forth their leaves. The oaks were beginning to burst their buds.

The plants in the meadows around Ekesjö were the following: *Millefolium, Alchimilla vulg., Viola martia inodora, Orobus tuberosus, Lathyrus pratensis, Equisetum arvense, Tormentilla, Veronica Pseudo-chamaedrys, Pilosella uniflora, Primula lutea, Anemone nemorosa, Hieracium, Pulmonaria gallica, Chaerophyllum sylvestre, Cerastium, Trifolium pratense album & rubrum, Juncus capitulis psyllii, Gnaphalium vulgare, Ulmaria, Laserpitium, Galium album quadrifolium, Leucanthemum, Silene viscaria, Galium luteum, Plantago lanceolata, Anthoxanthum, Quinquefolium minus procumbens* [*Achillea millefolium* L., *Alchemilla vulgaris* L. aggr., *Viola canina* L., *Lathyrus tuberosus* L., *Lathyrus pratensis* L., *Equisetum arvense* L., *Potentilla erecta* (L.) Räusch., *Veronica chamaedrys* L., *Hieracium pilosella* L., *Primula veris* L., *Anemone nemorosa* L., *Hieracium* sp., *Hieracium sylvaticum* L., *Anthriscus sylvestris* (L.) Hoffm., *Cerastium vulgatum* L., *Trifolium repens* L., *Trifolium pratense* L., *Juncus squarrosus* L., *Antennaria dioica* L., *Filipendula ulmaria* (L.) *Maxim.*, *Laserpitium latifolium* L., *Chrysanthemum leucanthemum* L., *Viscaria vulgaris* Bernh., *Galium verum* L., *Plantago lanceolata* L., *Anthoxanthum odoratum* L., *Potentilla verna* L.].

Viola martia [*V. canina* L.] had *stipulas foliaceas* [leafy stipules] at the root.

Ajuga tetragono-pyramidalis [*A.* 475 *erecta* of Index = *A. pyramidalis* L.], a kind of *Bugula,* occurred here and there along the road; it is not the same as the one commonly growing in Europe; the whole plant is one finger high, with an unbranched quadrangular spike; the leaves at the root are larger, ovate, hairy (*hispida*) but those on the stalk lying like scales one after another, curved upwards, reddish, hollowed below, with curly tips; there were three stalkless flowers at each of the lower leaves, but only one at the topmost. The calyx was inflated, five-toothed and downy. The upper lip was very short, bent and slightly notched.

"Kleran", as the little thrushes [*Turdus iliacus* L.] are called by Smålanders, cried and made music everywhere in the woods.

Ephemera nigra alis inferioribus albis [*E. vespertina* L.] was seen today here and there near to the water; we saw the same yesterday at Säthälla in such numbers that we have never seen anything like it.

In Skjutsemo there were some stone slabs on edge like runic stones.

After 1½ miles we arrived at Brånsmåla, where "skäfte" [*Equisetum hyemale* L.] grew in abundance.

Bearberry [*Arctostaphylos uva-ursi* (L.) Sprengel] grew along the way in full flower with clusters bent towards the ground.

After 1½ miles' journey we came to Vitlanda (Hwittlanda) in the evening. We rested there from 11 p.m. till 2 a.m.

The crickets chirped in Vitlanda so much that one could hear nothing else; the inhabitants entreated us not to hurt them, otherwise they would devour the clothes in the place where we had stayed overnight.

May 23 [p. 21]

At 2 a.m. we were up and on horse-back; we followed the main road eastwards to visit Småland's gold mine 1¾ miles from Vetlanda.

The weather was quite nice, everything was calm and silent. Mist had settled like little clouds on the bogs and could not rise, and the tender grass was dewy. The black grouse cooed far away, the thrushes were singing and each little bird warbled in its own tongue. The road was good and not hilly. When we approached Alsheda church, the sun rose through a thin cloud.

We passed the copper works at Ädelfors at 5 a.m. Here we took a quick look at the different ores, the roasters, the smelting furnaces, the refining furnaces and the washery.

Ore was of two kinds; one was a beautiful copper pyrites which from appearance alone was evidently a high-grade ore. The other ore is rare in Sweden and is a copper green (*Ochra viridis cupri*) in a grey limestone gangue with red patches and streaks; when this stone was broken, native copper was often found. Pink garnets were also found, especially in the stones which contained the native copper.

The ore washing plant was spacious and nicely constructed; maybe it will also prove useful in proportion to its size.

The roaster was a reverberatory furnace of German type.

The smelting furnace was a recent construction of a type uncommon in Sweden; but time will show whether it is of any advantage. Slags and blister copper were smelted together in the furnace, and the copper was supposed to sink to the bottom while the slag remained at the surface, and the laborious separation thus be avoided.

The junipers around the copper works looked like trimmed cypresses or *Cupressus meta in fastigium convoluta*; this is often seen around blast furnaces, brick works and other workshops where there is much smoke; if anyone could find the cause of this transformation of the junipers, this would be a great discovery for gardeners; if a gardener could produce such junipers with smoke only, cypresses would decrease in cost.

The workers and the inhabitants of Ädelfors were very dissatisfied and talked about all the contrivances with spite and disgust, foreseeing disasters.

The farmers complained that the parish, Alsheda, was too densely populated, but we assured them that on our way from Stockholm to here we never saw a more fertile soil than this; between the hills were the most delightful fields that could be imagined, without stone or gravel; the soil was sandy but with much mould and the only thing needed was manure to increase the amount of black earth; the many bogs in the neighbourhood were impossible to drain, but the mud could be used instead of dung.

A kind of rock which shattered like a slate, although coarser than the ordinary slate, occurred on the hills, also a kind of a rock which indicates that there may be ore in the vicinity.

We arrived at the gold mine at 6 a.m. The assayer, Mr Colling, showed us the small mineshafts, some of them had been followed with obvious eagerness while others were abandoned. The gold vein went *ad angulum acutum* downwards into the rock. The genuine gold was found in a pale pyrite. The lode was dark grey with some reddish streaks. The mountain slopes were nowhere steep and the small mineshafts started at the summit. Another mountain, separated from the gold mine by a large bottomless bog, had also yielded gold grains; so many discoveries of genuine gold in so many places so far away from each other do certainly indicate that an immense treasure is hidden here.

An adder [*Vipera berus* (L.)] lay on the road on our way back; it was stout and grey; its back had a dark zig-zag band and blue spots between its angles on the sides; the head was separated from the body by a blue *angulus acutus,* with white spots on the edge of its *fibra superior.* In the mouth on each side in the upper jaw there was a fang which could be erected and retracted like a cat's claw, indicating that this snake belongs to the genus which gives dangerous and often deadly bites.

A runic stone stood in the middle of the road ¼ mile from Vitlanda, but it was covered by a *Lichene crustaceo & leproso* and almost illegible, unless we had spent the whole day cleaning it; we discerned the following letters:

.... ᚼᚾ ᚱᚢ:ᚠ ....

...ᚱᚢᚦᚱ:ᛋᛒ·ᚾ:ᚦᚢᚱᛒᚢᚿ:ᛋᚢᚴ:ᚾᚠᚾ ....

Linnaeus: . . . *haru : f . . . ruþr : sb . a : þurbun : suk : afa. . . .*
Sm 110: . *arinum : sati stin : þansi : eftiʀ : heru : fþur : s——— . bruþr : sina : þurbun : auk : nfa :*

"Arinmund set up this stone in memory of Häre, his father (and) his brothers Torbjörn and Näve."

After 3½ miles' journey we came back to Vitlanda at noon, bathed in sweat in the intense heat.

There was a runic stone in the churchyard, but it was so badly knocked about that not a single word could be read. There was another runic stone near the south door of the church on which the following letters were seen:

ᛁᛁᛂ•ᚱᛁᛋᛐᛁ•ᛋᛐᛂᛂᚿ•ᚾᚠᛐᛁᛋᛘᚿᛒᚱᚮᚦᚢᚱ'

ᛐᚼᚼᚾᚱᛋᛐᛂᚿᛋᛋᛁ··

Linnaeus: *iie* × *risti* × *steen* × *aftismnbroþur– ahharstenssi. . . .*
Sm 109: *katil* × *risti* × *sten* × *efti sin broþur : nahhar stens su– +*

"Kettil raised the stone in memory of his brother . . . Sten's son."
[This runestone is now missing from Vitlanda Church.]

The many ruins in Vitlanda of walls, cellars, streets, etc. are said to be the remnants of the big town Wetala, which the centuries have transformed into a little village.

A dark soil which was said to turn blue in rain was seen along the road on Brankullen in Korsberga parish; there were small black granular stones in it which consisted of *mica.*

We travelled fast all night from Vitlanda to Stockatorp (1⅞), then to Nöberlöv (1⅝), and on the way we passed the Yxhult hills (⅛ mile). To Åshult 1 mile. To Åreda 1 mile. Here we began to see beautiful beech forests. To Växjö, 1½ miles, and we arrived there at 8 p.m.

May 24 [p. 25]

Växjö town stands almost in the centre of the land, is rather densely populated, with houses made of wood, the streets straight and light, recently planted with trees; it is not very large. Here is the seat of the district governor and chancellery, the bishop's residence and consistory, a high school and a school, but no factories except one for tobacco. St. Sigfrid's beautiful chapel was in a deplorable state, for it was all destroyed by fire a year ago, when a flash of lightning struck its high tower. We saw St. Sigfrid's clothes, his chasubles and mitres (*mitellas*) cloven at the top like a cardinal's hat. At Östregård we saw St. Sigfrid's fountain, where he is supposed to have baptized. Not far away were some stones with seven sockets bored into them, which are supposed to have held the candles during the baptism.

May 25 [p. 25]

The pharmacy in Växjö, directed by the chemist Falk, was so well equipped that it has few equals in the country.

The school was visited at 9 a.m.; there were 80 pupils in the upper school and about 150 in the lower school.

The Solberg, which stands north of the town, has a rusty-coloured shattered rock in many places and also in one place pyrites, which both indicated that the rock might contain iron ore. On the mountain, bird cherry, blackthorn, bramble, lily-of-the-valley, *Turritis major* [*T. glabra* L.], *Saxifraga alba* [*S. granulata* L.], *Orobus tuberosus* [*Lathyrus tuberosus* L.], *Paris* [*P. quadrifolia* L.] and cudweed were in flower. On the east side of it there was a suicidal cliff called "Dödsprång" (Death-leap).

Hofslund was still further north, a very nice grove with beech, birch, oak and other deciduous trees.

Dermestes testaceus pilosus, elytris striatis retusis praemorsodentatis [*Ips typographus* (L.)] was found in great numbers in the timber logs. The zoologists have long wanted to know what insect produces the powder between bark and wood in timber and fences, also cuts figures, like letters, in the wood; here we had the opportunity to see it.

Curculio subfuscus, elytris fasciis duabus testaceis [*Hylobius abietis* L.] was also found in Hofslund; it clings more firmly to the finger than any other insect; this is due to a rigid claw at the end of the leg, besides the two ordinary ones on each foot.

May 26 [p. 26]

Assessor Johan Rothman, district doctor, lector of natural science and a man of much learning and experience, showed us a strange mushroom or *Phallus volvatus, pilei apice clauso* [*P.* 1095 *peniformis* of Index = *P. impudicus* L.] found in the neighbourhood of Växjö and so rare in Europe that few botanists have seen it; I asked him to send it to the Royal Academy of Science, which he did. See the *Handlingar,* 1742, p. 19.t.2.f.1. ["En sälsam swamp funnen in Småland", by Johan Rothman.]

Eventually we left Växjö; we saw stone fences recently erected, but only a few, since the inhabitants are unaccustomed to them; one is not anxious about the woods here and it is more laborious to build in stone.

Sapwood (Sava) was being carried home by boys, who met us. Sapwood is so common in Sweden that every child has eaten it, but abroad it is as rare as it is common here; hence it will be shortly described. The time when the pine tree has put forth its annual shoots to a length of ½-1 finger is the right time for collecting sapwood. The bark is cut at the branches and then peeled off like skin from the trunk. This is easily done at this time of the year; when the internode is bared, the sapwood surface beneath has a sweet taste and is soft like jelly. This pulp is removed from the trunk with a knife or a thin wire of brass or steel. The sapwood is then folded up and eaten fresh without any preparation. If it is left for any longer it turns tough, resinuous and unsavoury. The tree from which the sapwood has been collected is cut down and used as fuel for the stove during the winter. The trees always dry up and die above the zone where the bark is removed. This sapwood is a veritable balm or resin when dissolved in water. It is collected and eaten by boys and youths for its good taste, but it is also a remedy for exanthemata, worms, shortness of breath, consumption and scurvy. It is a powerful diuretic without being harmful and is less irritating to the stomach than any other balm, without being a purgative.

We had a 1⅝ mile journey to Åreda.

Nightjars [*Caprimulgus europaeus* L.] flew in front of us here and there on the road, when the sun had set and the night was dark.

We had two more miles to ride to Lenhovda, where we spent the night.

May 27 [p. 28]

Rain, thunder and lightning forced us to remain in Lenhovda till 3 o'clock in the afternoon.

Last winter, 1740-1741, was neither so hard nor so long as the winter of 1739-1740, and yet it produced three times as much damage, especially here in Småland, where it was worse than in Uppland; not only apple trees and pear trees, but plums and cherries and even currants were destroyed; this was due to the snow that fell before Michaelmas when the trees were still green and thus still retained all their juices.

Superstitions, which are very tenacious, especially with the uneducated and in the far-off provinces, were here, many of them left from popish and heathen times. From the latter period people retain the idea that when one farmer's cow dies, the farmer can transfer the bad luck to his neighbour's cattle if he buries the carcase in the neighbour's field or dung hill. *Febres Ephemeres* are treated according to the following method, which emanates from popish times; when someone starts to shiver and then feels hot, and the body feels all bruised, and there is a nose bleed or blood spitting, and this goes on for a few days, the simple man believes he is being ghost-ridden and the disease is cured by reciting the following verse:

Christ Our Lord and Saint Peter
walked along the road
There they met a dead man.
Saint Peter and Our Lord
asked the dead man
Where are you going?
The dead man said:
I am going to N.N. [Jonas, etc.]
What for?
Said Our Lord and Saint Peter.
I will squeeze his heart blood out of him.
No, I will not let you do that,
I will bury you under logs and stones
and you will harm no man.

Dye-plants, which within the country, are often used by the country folk. "Black earth", an earth, is found here in Lenhovda parish at Signelstorp in a wood ¼ mile from the church.

A similar earth occurs near Horshult in Wirestad parish and is said to be a kind of mud that gives a dye blacker than anything else. Another man said that the cloth is dyed black with an earth from Boderups parish.

Red is obtained here in the parish from *Byttelet* [*Ochrolechia tartarea* (L.) Massal.], which is brought from Västergötland and looks like a dark clay spotted with red. For use it is boiled with water and a little urine.

Yellow is obtained in several ways. Some boil the yarn with birch leaves. Others use roof moss. This roof moss is a *Lichen orbiculis peltatis* [*Lepraria flava* (Schreb.) Smith = *L. candelaris* (L.) Fries] of yellow colour which grows on old shingle roofs, preferably on churches. The moss is boiled in water with or without alum. Both methods give a rather intense yellow colour, paler if alum is used. Others dye their yarn yellow with bark from hornbeam (*Carpinus*) which is pounded and boiled in water, but this colour is not as intense as the one produced with buckthorn [*Rhamnus catharticus* L.].

Green is obtained here from the berries of alder buckthorn (frangulae) [*Frangula alnus* Miller]. After the yarn has been dyed yellow with birch leaves and dried, it is boiled once more with these berries, which yield a green colour.

Brown is obtained from "stone-moss" [*Parmelia saxatilis* (L.) Ach.], which is boiled in water. The moss must be strained off before the yarn is added, otherwise it will be unevenly dyed. Soot boiled in beer yields a brown colour too, but paler than the stone moss.

Most "stone mosses" [lichens] are dye-yielding. The violet moss [lichen] which gives the red colour of the violet stone yields an intense yellow colour if it is rubbed with a glove or wrapped in a wet cloth. It is remarkable that the women here are not as particular as on Åland about collecting single moss [lichen] species only, but they pick all kinds which grow on stones.

When the farmers here make candles they often colour them with a yellow moss (*Lichen leprosus luteus*) [*Lepraria candelaris* (L.) Fries] to make them the colour of wax. This moss looks like a yellow mould and grows on old walls. It is scraped off, wrapped in linen cloth and boiled in water which turns yellow. This yellow water is used to dilute the tallow used for the dipping. Poor people mix the tallow with much spruce resin. This gives more and bigger candles, but these cheap candles burn fast with a big flame, and they drip, smell and smoke.

We had 1 mile to go from Lenhovda to Närhult.

Andromeda [*A. polifolia* L.] with big pink flowers grew in the marshes on both sides of the road.

All the day we saw tall heather growing along the road. It contributes greatly to the maintenance of the bees, which are common here. The fields were covered with pebbles, which were supposed to be of no harm to the crops and retain the moisture in the soil.

Rotten oaks lay on the ground; they would have served better as manure for lovely lilies and bulbs, since there is no better fertilizer for these plants; they can also be used for garden paths to keep them free from weeds.

Two miles to Villeköl. The inn was empty. There was no innkeeper and no ostler; thus we had to continue on exhausted horses to the next inn; an inconvenience to travellers which I have never come across before in this country.

One and a half miles to Brännhult in the county of Kalmar.

Oaks are the most useful trees in the country, especially for ship building. People commonly complain that the oaks grown in Sweden are more branched and curved than the foreign ones. People have tried to make them grow straight by frequent lopping but with little success. In this place we saw straighter and more uniform oaks than we have ever seen before; consequently we wanted to know why; then we found out that the nature of the soil is of no importance, only the oaks should grow close enough to each other, in the same way as pine trees growing in a dense forest become slender and less branched, as if they were competing for space high up; the lowest trees are thus suffocated and the higher develop a stouter trunk. Those who want to plant oaks and make them grow straight should thus crowd them together at first and clear out the smaller ones later, which will make them grow straight. The same happening can be seen in the pine forests, where the lankiest bean-poles after some time yield the biggest and straightest logs and masts, and they never develop thus if they are allowed to spread out their exuberant branches when young.

The woods at ¼ mile along the road to Villeköl had been destroyed by a forest fire caused by burn-beating.

The flowers seen today were ash [*Fraxinus excelsior* L.], bearberries [*Arctostaphylos uva-ursi* (L.) Sprengel], marsh rosemary [*Ledum palustre* L.] *Veronica foemina* [*V. serpyllifolia* L.], *Trientalis* [*T. europaea* L.] and *Hottonia* [*H. palustris* L.].

Bearberries grew in the most infertile gravel, where no other plant can nourish itself, and remarkably enough they seem to thrive better here than in a more fertile soil.

Yews (*Taxus*) were said to grow in Kungshärad near Lindfors bruk in Ryd parish. The inhabitants boil the yew wood in water and wash themselves with the decoction as a treatment against scabies. Now that we had reached the Kalmar district, the landscape changed and was softer and more pleasant. We saw a beech forest today.

One and a quarter mile to Bergsryd.

One and a quarter mile to Hårby.

One and a quarter mile to Kalmar, where we arrived at 6 a.m. after riding all night.

May 28 [p. 32]

At a distance Kalmar town looked magnificent with its castle, its moats, redoubts and other fortifications; its position at the seaside, the delightful church and its stone houses make it look very pleasant. There was no fresh water in the town. There was a well in almost every cross-street, but the water was always salt. We were told that the only fresh water supply was a well in the castle. There was a smell of seaweed in this town. The walls were yellow since the stones were covered with a *Lichene fulvo sinibus daedaleis laciniato* [*Xanthoria parietina* (L.) Th. Fries]. The cathedral is in the centre of the town; it is built like a cross with a quadrangular corner in each *sinu* [bay] and has no high towers; the inside is beautiful, and the chasubles are very fine. Our greatest difficulty here was to find a place where we could eat. We searched for several hours until we found it at the apothecary, more for our supplications than for our money. Cellar and wines were in accordance therewith.

We saw the slaves or prisoners who are sentenced to work here at the prison for major crimes, toiling like horses; at 6 p.m. they are driven into their dark dens under the walls. They have to manage on four *stiver* a day, but if they work from 6 a.m. till the evening with a break from 11 to 1, they receive six *stiver* a day. Their distress made our hair stand on an end; there is no escape for them; they are doomed to prison for the rest of their life.

Two forts, Grimskär and Käringlåret, were built ¼ mile out in the sea on two little islands. The first one was in the channel where the ships pass in front of the castle. There was said to be a fresh water well there. The other one was closer to the embankment.

The corners of the houses in the town were fitted with boards to make them withstand weather and the wind's force.

Fucus folio dichotomo integro, caule medium folium transcurrente, vesiculis verrucosis terminatricibus [*F.* 1002 *Quercus marina* of Index = *F. vesiculosus* L.], the seaweed which was lying on the shore and growing under water, had two kinds of bladders. One kind was smooth, without grains, with a material which looked like spider's web inside the cavity; the other kind of bladder at the tips of the seaweed was oblong and usually occurred in groups of three. There were granules in their cortex and a white pulp in the cavity.

Ulva prima [*Enteromorpha intestinalis* (L.) Link] was thrown up on the shore, all green and hollow like inflated intestines.

Arenea oblongus, fuscus, abdomine nigro, apice spinoso, tergo quatuor punctis depressis notato [*Steatoda quadripunctata* (L.)] was found in the houses; it had dark brown feet, a soot-coloured breast, a black belly with grey hair and 5-6 pale teeth at the rear end whence the threads are spun.

The weather today was cloudy and chilly.

May 29 [p. 34]

There was such a strong wind today that we did not dare to cross to Öland. We spent the day looking at the town, the castle and the dye works. The dye works are outside the town because dyeing cannot be done in the town, where the water is salty.

Saw-wort (*Serratula*) [*S. tinctoria* L.] was bought from Öland at a price of ten *stiver* a pound. It grows sparsely around Kalmar and yields a yellow dye.

Safflowers (*Carthamus*) [*C. tinctorius* L.] are flowers bought abroad and used for dyeing rosy red. This thistle could well be cultivated in southern Sweden to the benefit of the dye works; it should be noted that a rich soil will make it produce very few flowers while a dry soil yields more but smaller flowers.

The ordnance store was seen in the morning, with fire-arms and gun mountings and other *tormentis bellis* [military instruments of torture], a martial collection with swords, pistols and guns nicely arranged as in a library; there were also workshops for smiths, joiners and turners close to the ordnance store.

The streets in the town were straight and most of them ended in a gateway on the embankment surrounding the town, often leading to a little jetty.

We left the ordnance store and the town and walked southwards towards the castle, which was in front of us with the sea to our left and a field to our right. On the field grew almost nothing but houndstongue [*Cynoglossum officinale* L.] and the plant which is called "Mannablod" [Man's blood].

This mannablod or manna-ort [man's herb] is a plant which is much talked about in Sweden. Several members of the Parliament especially asked me to find out what herb it might be, for it was said that it grows in no other place in the world but here at Kalmar castle, where it once grew up from the blood of Swedes and Danes, killed in warfare on this field. We were certainly very intrigued by so remarkable a thing in nature, since it is the practice of nature to procreate and maintain what is created rather than to create anything new; we were much taken aback when we realized that the plant was nothing but the common *Ebulus* or *Sambucus herbacea* [*Sambucus* 251 *Ebulus* of Index = *S. ebulus* L.], which grows wild in the greater part of Germany, around Växjö and in gardens, where it multiplies so quickly that it is difficult to eradicate. However its usefulness is greater than its rarity, especially as it grows here in such abundance; one could annually supply all pharmacists with *Ebulus* roots, leaves, flowers, berries, seeds, pulp and bark, which are all used in pharmacies. Some people have tried the berries as a dyeing agent, and told us they they yielded a pleasant violet colour.

We visited the castle and looked at its walls, which have endured so much gun-fire, and walked around the embankments.

May 30 [p. 35]

This was a cloudy day, cold and unpleasant with a strong wind; we could hardly go outside the door, far less cross to Öland.

The trade of this town is based on tar, boards, oak wood, flag stones, potash and alum from Lovers' alum works.

May 31 [p. 36]

It was equally impossible for us to cross over to Öland today; storm, hail, clouds and cold were just as intense today as yesterday.

We went out to botanize with the pharmacist, Mr Nordstedt, making our way to the Royal Stables ¼ mile from the town, which are surrounded by vast meadows and beautiful oak groves.

Scandix seminibus hispidis [*S.* 242 *echinata* of Index = *S. anthriscus* L. = *Anthriscus caucalis* Bieb.], which is so similar to the true chervil and grows on the dung-hills in Holland, has never before been seen in Sweden; it grew outside the batteries along the fences.

Ranunculus foliis rotundis et capillaceis [*R. aquatilis* L.] covered the ditches with snow-white flowers.

Hottonia [*H. palustris* L.] flourished in a small pool close to the farmyard.

Hydrocotyle foliis peltatis orbiculatis undique emarginatis [*H.* 221 *vulgaris* of Index = *H. vulgaris* L.], a little plant which has not been found in Sweden before, occurred abundantly around the small pool where the *Hottonia* grew.

Scorzonera caule subnudo unifloro, foliis nervosis planis [*S.* 647 *humilis* of Index = *S. humilis* L.] occurred all over the meadow as well as the farmyard in the dry places. This *Scorzonera* is very similar to the garden one, but is smaller and has longer roots. The basal leaves are a quarter ell long [15 cm], lanceolate (*lineari-lanceolate*), acute, even at both ends, with three veins. The stalk ¼ ell long, straight, grooved (*striatus*), brown at the top, slightly hairy and one-flowered, has two leaves on it, which are much narrower and smaller than the basal leaves. The involucral scales are more pointed than in the cultivated form. The flower-head is yellow and the florets are just as long as the involucral scales. The pappus of the seeds radiates from the sides like a feather. I have seen it in Skåne and Småland, but since it does not grow in the northern provinces, it has not been described among the Swedish plants before. The *Radix* one finds in pharmacies is *radix Scorzonerae,* which in most parts of Europe comes from the garden *Scorzonera* but unfortunately so, since cultivated plants always have a far milder taste than wild ones, as can be seen in *Lactuca, Cichorium, Asparagus,* etc., thus losing the strong medicinal properties for which one seeks in pharmacies. Hence the roots of wild plants should be collected for the pharmacies, the garden *Scorzonera* used in cooking. Physicians often complain that *radices Scorzonerae* do not exert the strong sudorific and other effects described in the books, but the cause is that it is the root weakened by cultivation which is taken in pharmacies.

The common plants at this time of the year were: *Paris* [*P. quadrifolia* L.] ; *Pulsatilla 1 : a* [*Pulsatilla vulgaris* Miller] ; *Ranunculus foliis radicalibus reniformibus crenatis, caulinis digitatis linearibus* [*Ranunculus auricomus* L.] ; *Anthyllis* [*A. vulneraria* L.] ; *Plantago foliis lanceolatis, spica subovata* [*P. lanceolata* L.] ; *Turritis foliis omnibus dentatis hispidis, caulinis amplexicaulibus* [*Arabis hirsuta* (L.) Scop.] ; *Arabis foliis lanceolatis petiolatis integerrimis* [*Arabidopsis thaliana* (L.) Heynh.] ; *Cardamine 1 : a* [*C. pratensis* L.] ; *Geum 1, 2* [*G. urbanum* L., *G. rivale* L.] ; *Trientalis* [*T. europaea* L.] . *Pinguicula 1* [*P. vulgaris* L.] was in full flower. *Viola caulibus adscendentibus floriferis, foliis cordatis* [*V. canina* L.] was at its most beautiful. *Orchis palmata* [*Dactylorhiza* sp.] had no flowers yet. *Serratula* [*S. tinctoria* L.] had no stalk. *Primula foliis crenatis glabris, limbo florum plano* [*P. farinosa* L.] was abundant. *Juniperus mas et foemina* [*J. communis* L., male and female] were in flower. *Agaricus minimus capitulo turbinato plano albo, lamellis margine fuscis* [*Agaricus umbellifer* L. = *Omphalina ericetorum* (Fries) M. Lange] was found in one place only.

We did not find many insects today, only *Chrysomela thorace viridi caeruleo, elytris rubris apice nigris* [*Lina populi* (L.)] ; *Cantharis elytris nigricantibus, thorace rubro nigra macula* [*C. fusca* L.] ; *Cicindela viridi-aenea, elytris punctis latis excavatis* [*C. riparia* L.] .

Outside the town we saw the well where the inhabitants get their fresh water, which is expensive due to long and difficult transport.

June 1 [p. 38]

It was almost as windy today. We were tired of Kalmar, after having waited four days for better weather; thus we decided to cross on the ferry-boat and pass the many shoals between Kalmar and Öland. Spotted fever, which afflicted two of the inhabitants of the house where we had stayed, the shortness of summer and the pleasant prospect of Öland hurried our departure. At 3 p.m. we embarked on the ferry in a gale from the south-west, leaving Kalmar and more than 30 small vessels behind us.

ÖLAND

June 1 [p. 39]

As soon as we touched the shore of Öland we realized that this was a land which was altogether different from the rest of the Swedish provinces. Thus we decided to annotate everything we would see on this island all the more meticuously.

The plants seen today were the following:

Veronica mas 3 [*V. officinalis* L.]
Femina 9 [*V. serpyllifolia* L.]
Pseudochamaedrys 7 [*V. chamaedrys* L.]
Anthoxanthon [*Anthoxanthum odoratum* L.]
Poa 4 [*P. pratensis* L.]
Valeriana 1 [*V. officinalis* L.]
Alchemilla 1 [*A. vulgaris* L. aggr.]
Galium luteum 1 [*G. verum* L.]
Album [*G. boreale* L.]
Aparine minor 2 [*Galium uliginosum* L.]
Anchusa 1 [*A. officinalis* L.]
Cynoglossum 1 [*C. officinale* L.]
Primula lutea 1 [*P. veris* L.]
Rubra 2 [*P. farinosa* L.]
Campanula 1 [*C. rotundifolia* L.]
Verbascum nigrum 2 [*V. nigrum* L.]
Pimpinella [*P. saxifraga* L.]
Selinum 1 [*Peucedanum palustre* (L.) Moench]
Chaerophyllum 1 [*Anthriscus sylvestris* (L.) Hoffm.]
Juncus capit. psyllii 11 [*Luzula campestris* (L.) DC.]
Nemorosus 10 [*Luzula pilosa* (L.) Willd.]
Rumex arvensis 4 [*R. acetosella* L.]
Epilobium montanum 3 [*E. montanum* L.]
Saxifraga alba 1 [*S. granulata* L.]
Arenaria serpillifolia [*A. serpyllifolia* L.]
Cerastium viscosum [*C. vulgatum* L.]
Agrimonia [*A. eupatoria* L.]
Prunus sylvestris [*P. spinosa* L.]
Fragaria [*F. vesca* L.]
Rosa [*R. canina* L.]
Rubus fr. caesio 3 [*R. caesius* L.]
Potentilla 1 [*P. anserina* L.]
Geum rivale 2 [*G. rivale* L.]
Tormentilla [*Potentilla erecta* (L.) Räusch.]
Anemone nemorosa 2 [*A. nemorosa* L.]
Ranunculus nemorosus 6 [*R. auricomus* L.]
Acris 10 [*R. acris* L.]
Bulbosus 13 [*R. bulbosus* L.]
Fol. capillaceis 16 [*R. aquatilis* L.]
Ficaria 4 [*R. ficaria* L.]
Caltha [*C. palustris* L.]
Ajuga [*A. pyramidalis* L.]
Glechoma [*G. hederacea* L.]
Rhinanthus 2 [*R. major* L.]
Draba 1 [*Erophila verna* (L.) Chevall.]
Cardamine 1 [*C. pratensis* L.]
Turritis hirsuta 2 [*Arabis hirsuta* (L.) Scop.]
Geranium robertianum 7 [*G. robertianum* L.]
Polygala alba 1 [*P. vulgaris* L.]
Anthyllis [*A. vulneraria* L.]
Orobus tuberosus [*Lathyrus tuberosus* L.]
Vicia sativa [*V. sativa* L.]
Lathyrus pratensis 2 [*L. pratensis* L.]
Lotus 1 [*L. corniculatus* L.]
Ononis inermis 1 [*O. repens* L.]
Hypericum caule quadrangulo 1 [*H. maculatum* Crantz]
Gnaphalium mont. 1 [*Antennaria dioica* (L.) Gaertner]
Chrysanthemum Leucanthemum 2 [*C. leucanthemum* L.]
Achillea millefolium 1 [*A. millefolium* L.]
Viola inodora 2 [prob. *V. canina* L.]
Betula 1 [*B. verrucosa* Ehrh.]
Juniperus mas & femina [*J. communis* L.]
Equisetum arvense 1 [*E. arvense* L.]
Polypodium, Filix mas 2 [*Dryopteris filix-mas* (L.) Schott]
Filicula 5 [*Cystopteris fragilis* (L.) Bernh.]
Polytrichum vulgare 1 [*P. commune* L.]

Ranunculus radice bulbosa 13 [*R.* 469 *bulbosus* of Index = *R. bulbosus* L.] was rather hairy; the flower-stem was five-angled and the sepals were bent back; whereby it can be distinguished from *Ranunculus acris 10* [*R.* 466 *acris* = *R. acris* L.], which is twice as tall with rounded flower-stems and sepals which are not bent back.

The blackthorn bushes were made white by a moss [lichen], which almost entirely covered them, *Lichenoides LV Dillenii* [*Evernia prunastri* (L.) Ach.].

Alsine foliis lanceolatis serrulatis [*A.* 371 *grandiflora* of Index = *Stellaria holostea* L.], which has not been described as a Swedish plant before, was found growing under the junipers.

The wood was full of birch and juniper, among which grew blackthorn and dog roses, so that only with the greatest difficulty could one get through it. The soil was a deep and beautiful black mould, disturbed everywhere by the swine, and only clearing and fencing were needed to turn it into a lovely meadow.

There were many fossils or "Öland spikes", *Helmintolithus nautili recti* [orthoceratite], in the stones in the walls.

A strangely shaped petrifaction, which was common in this place, we considered to be an impression of a *Helmintolithus echini*; only one side was seen of its shell, which had been preserved in the stone.

Midges in thousands, like enormous swarms of bees, flew from the bushes near the shore; one would have imagined oneself in the wildest part of Lapland, if these gnats had been blood-sucking; we have never before seen so many midges. This insect can be called *Tipula thorace virescente, alis membranacei coloris, puncto nigro* [*Chironomus plumosus* (L.)] . This midge is twice as big as the blood-sucking mosquito. The wings lie flat, are white with a black spot in the middle towards the outer edge formed by a pair of veins which there anastomose. The body is black, slightly hairy, lighter at the waist; under the breast there is a big pouch. The midge opens it mouth and moves it but does not suck blood. The forefeet are longer than the second pair. The male's antennae are bristly with forward pointing hairs. The tail is curved upwards. The female is similar to the male, but its abdomen is only a little longer than the wings. It does not raise its tail as high and its antennae are much less hairy.

Scarabs were rolling in the horse dung, among them being a *Scarabaeus magnus niger vulgatissimus, antennis articulatis. Raj.ins. 74 n.1.* [*Geotrupes* or *Scarabaeus stercorarius* L.] which was intensely blue and somewhat small. *Scarabaeus capite thoraceque atro opaco elytris cinereis nigro-nebulosis* [*Ontophagus nuchicornis* (L.)] was completely black and oval, except the elytra which were shorter than the abdomen and scarcely grooved. *Scarabaeus capite thoraceque nigro, glabro, elytris griseis, pedibus pallidis* [*Aphodius prodromus* (Brahm)] is much smaller but more oblong. The wing cases are almost as long as the abdomen, grey, scarcely grooved, with a longitudinal oblong pale marking along the back which bifurcates at the head end.

Hirundo dorso nigro-caerulescente, rectricibus immaculatis [*Martella urbica* (L.), house-martin] is the swallow which builds nests outside the houses under the eaves. The bill is black, short, broad and almost triangular. The entire body is black on the upper side, but with a blue hue on head and back. Nostrils bare, eyes black, wings rather long, darkish, as long as the tail when the bird is sitting. There are 18 pinions, of which the first 9 are pointed and one proportionally longer than the others, tapering inwards, also the outermost the longest, but the inner 9 are of equal length, blunt and cloven at the tip. The tail feathers are 12, of which the outer ones are longer and the outermost the longest. All these feathers with their upper coverts are dark. The under parts–chin, breast, belly, under tail and wing coverts and the rear end of the back–were white. Legs and toes had a white wool on them. The claws were small and all the toes were of equal length.

Cancer, *Pulex fluviatilis dictus* [*Gammarus pulex* (L.)] , was found in the gravel on the seashore in the afternoon. It was much smaller than a shrimp and little bigger than a midge. The body was compressed and had on its forehead four horns of almost equal size, which consisted of three rigid joints and an outer part which was like a hair with an interminable number of joints; it had six pairs of feet. The first pair had chelas and a moveable thumb at the tip, but nothing to set against the thumb. The next pair of feet was attached to the root of the tail and were curved forwards under the abdomen, the last pair were cloven at the tip (*bidigitati*) and were attached under the tail. The tail was pointed with joints with an oblong red spot at the outer margin. This can be told from other crayfish, in that it has no stout cuirass on the thorax and all the joints of the back are equally big.

Three kinds of mussels were found on the shores.

Concha testa subrotunda, sulcis 26 longitudinalis, tribus transversalibus [*Cardium edule* L.]. Its shell was white with 24-26 deep furrows with raised stripes between. On the inside the shell was also furrowed. Its basis looked as if it was covered with two small shells on each side, the outer smaller than the inner. Each of the shells was rounded and as big as a lupin seed.

Concha subrotunda glabra incarnata [*Tellina balthica* L.] was almost rounded, somewhat triangular, less bulging, thin, flesh-coloured, smooth with a few lines hardly visible to the naked eye, and smaller than the first one.

Concha mytulus dicta [*Mytilus edulis* L.] was black, with a violet hue, oblong like a ham, with a blunt tip, compressed on one side, smooth and with stripes going around like a horse-shoe.

We spent this night in Färjestaden.

June 2 [p. 43]

We left Färjestaden at 4 o'clock in the morning. The pharmacist from Kalmar, Mr Nordstedt, accompanied us in order to look for medicinal plants. We rode towards Borgholm castle. The weather was fine; the ground was even; the thrushes sang in the trees.

The fences were of more different kinds and materials than we had seen in any other place; thus

- α. Dry stone walls of a slate-like red limestone; the stones were simply laid on top of each other to a height of 1½ ells [90 cm], and armed on top with dry blackthorn twigs.
- β. Wattles of juniper, the twigs laid horizontally between perpendicular stakes.
- γ. Juniper shrubs laid sidewise and fastened to perpendicular juniper stakes.
- δ. Fir boards, split thin, often half an ell [30 cm] broad, laid obliquely.
- ε. The fences surrounding the houses and the farmyards were made of stout stakes notched at the top and transverse beams resting on the notches; inside were upright oak planks leaning on the transverse beams, so that fences were easily opened inwards by pushing the planks inwards but the cattle inside the yard could not get out.

The houses were built of oak boards morticed into oak posts, without knobs; with clay in the joints to stop up the chinks. There were quadrangular draught valves on top of the chimneys, which could be drawn up or let down when ever needed.

Ropes were made of linden bast without previous retting.

The soil in the strip of shore between the sea and the plain was a deep, pure and beautiful black mould. There were oak, birch, alder, linden, hazel, blackthorn and juniper in the woods.

We did not see a single heather plant during the whole day, only green turf everywhere.

A ¼ mile from Färjestaden we left the main road towards the left and rode on to a beautiful meadow belonging to Björnhovda. This district was very pleasant with many deciduous trees, especially birch and hazel. Here occur the rarest plants, such as have never been heard of in Sweden before, and to see which I travelled in 1738 from Paris to Fontainbleau, where I saw them without thinking I would ever see them again.

Cypripedium bulbis subrotundis, foliis oblongis caulinis [*Cypripedium* 737 *muscifer* of Index = *Ophrys insectifera* L.] is commonly called *Orchis muscam referens.* Its flowers bear such a resemblance to flies, that an uneducated person who sees them might well believe that two or three flies were sitting on the stalk. Nature has made a better imitation than any art could ever perform. *Radix testiculata. Caulis seminudus. Bracteae florae longiores. Perianthum triphyllum, patens, aequale. Petala tria, holosericea, purpurea, quorum duo erecta, linearia; tertium est labium inferius oblongum, trifidum; lacinia intermedia maxima bifida; ad basin labii puncta utrinque duo nigra; in medio labii macula cyanea.*

Orchis bulbis indivisis, nectarii labio quinquefido punctis scabro, cornu obtuso, petalis conniventibus [*O.* 725 *militaris* = *O. militaris* L.]. This magnificent flower has never been seen in Sweden before. It grows here abundantly and is generally called *Orchis militaris hiante cucullo major. Radix testiculata. Caulis palmaris. Petala conniventia & fere connata in galeam erectam, purpuream, striatam, externe albidam. Labium inferius angustum, purpurascens punctis holosericeis rubris, trifidum laciniis linearibus: intermedia productiore trifida: media minima. Nectarium brevissimum, obtusum.*

Orchis bulbis indivisis, nectarii labio quadrifido, punctis scabro, cornu obtuso, petalis distinctis [*O.* 726 *ustulata* of Index = *O. ustulata* L.]. This is generally called *Orchis militaris minima* and is as unknown in Sweden as the two others. *Radix testiculata. Caulis digiti longitudine. Spica subrotunda, compacta. Bracteae brevissimae. Petala quinque, conniventia in galeam purpuream. Labium inferius trifidum, album, antice punctis purpureis maculatum: intermedia productiore bifida. Nectarium brevissimum, obtusum.*

Orchis bulbis indivisis, nectarii labio quadrifido crenulato, cornu obtuso [*O.* 724 *morio* = *O. morio* L.], generally called *Orchis morio foemina,* although it is not unknown in other places, merits a description. *Radix testiculata, amplexicaulia. Spica rara obtusa. Bracteae germine breviores. Petala quinque, conniventia in galeam. Labium inferius serratum, lateribus retroflexum, trifidum: intermedia emarginata. Nectarium obtusissimum ascendens, germine brevius, saepius emarginatum. Variat flore purpureo lateribus flavis.*

Orchis bulbis palmatis, nectarii cornu setaceo germinibus longiore, labio crenato [*O.* 727 *longicalcar* of Index = *Gymnadenia conopsea* (L.) R.Br.] is generally called *Orchis palmata calcaribus oblongis.* It is rare in Sweden and merits description too. *Bracteae longae. Flos purpureus. Galea triphylla, connivens, foliolis calycinis lateralibus patentissimis. Labium inferius breve, trifidum, subaequale.*

Valeriana dioica or *Valeriana palustris minor* [*V.* 31 *dioica* of Index = *V. dioica* L.], I have only seen it on the plains of Skåne before. Both males and females grow in the groves here. *Caulis vix digiti longitudine. Folia pinnatifida. Maris corolla triplo major, staminibus tribus, pistillo nullo. Feminae flores minutissimi, magis conferti, pistillo triplici stigmate; staminibus quidem tribus minutissimis intra tubum corollae haerentibus & obsoletis sed antheris amnino sterilibus.*

Plants common in the meadows were *Mercurialis caule simplicissimo foliis scabris* [*M. perennis* L.], *Polygonatum latifolium vulgare* [*P. odoratum* (L.) Druce], *Opulus* [*Viburnum opulus* L.], *Xylosteum* [*Lonicera xylosteum* L.], *Centaurea 1* [*C. scabiosa* L.], *Clinopodium* [*C. vulgare* L.], *Geranium 1, 2* [*G. sanguineum* L., *G. sylvaticum* L.], *Thymus 2* [*Acinos arvensis* (Lam.) Dandy], *Rubeola Cynanchica* [*Galium triandrum* Hylander], *Briza* [*B. media* L.], *Trifolium* [*T. montanum* L.], *Linum catharticum* [*L. catharticum* L.], *Medicago 1* [*M. falcata* L.], *Pinguicula 1* [*P. vulgaris* L.], *Cardiaca* [*Leonurus cardiaca* L.], *Euphrasia 1* [*E. officinalis* L. aggr.], *Serratula 1* [*S. tinctoria* L.], *Arctium* [*Arctium* sp.], *Melampyrum* [prob. *M. pratense* L.], *Scorzonera* [*S. humilis* L.], *Alliaria* [*A. petiolata* (Bieb.) Cavara & Grande], *Campanula trachelium dicta* [*C. trachelium* L.]. *Papaver erraticum* [*P. dubium* L.], which is used in the pharmacies, grew in the fallow fields, and caraway in the headland. *Thlaspi arvense* [*T. arvense* L.] grew in the cornfields more abundantly than anywhere else, together with a little *Sinapis rapistrum dicta* [*Sinapis arvensis* L.].

Elder and service-tree grew wild and abundantly in the woods, so there is no need for the pharmacists to order elder juice from Germany.

Pulsatilla flore minore nigricante C.B. [*P.* 447 *retroflexa* of Index = *Pulsatilla pratensis* (L.) Miller], which I have seen before growing wild around Lübeck, but never in Sweden, occurred in all dry patches.

The winter rye had ears but was hardly one ell [60 cm] tall; the barley was magnificent.

A ¼ mile before one reached Isgärde, stood Glömminge church on the left. Everywhere the groves looked like gardens on Carlberg.

After 1¼ miles we changed horses at Isgärde inn; then we rode into a big forest of oak and pine trees, very good for timber.

Near Rälla Gård grew *Aegopodium* [*A. podagraria* L.], *Angelica sylvestris, Laserpitium* [*L. latifolium* L.], *Aster salicis glabro folio* [*I. salicina* L.], several orchids and *Euonymus* [*E. europaeus* L.], not seen in Sweden before except in Skåne.

Orchis alba bifolia calcari oblongo C.B. [*O.* 723 *Satyrium* of Index = *Platanthera bifolia* (L.) Rich.] was not in flower yet.

Orchis bulbis subpalmatis rectis, nectarii cornu conico: labio trilobo integerrimo, bracteis flore longioribus [*O.* 728 *sambucina* of Index = *O. sambucina* L.], which is generally called *Orchis palmata Sambuci odore,* varied with red, white or rusty red flowers. *Radix palmata, oblonga, minus digitata, recta, deorsum protensa. Folia immaculata. Bracteae longitudine corollae. Petala tria exteriora patentia, duo interiora conniventia in galeam. Labium inferius trifidum, subcrenatum, lacinia intermedia angustiore non vero breviore. Nectarium longitudine germinis, obtusum.*

Orchis bulbis palmatis patentibus, nectarii cornu geminibus breviore: labio crenato, dorsalibus patulis [*O.* 729 *basilica* = *O. maculata* L.] is generally called *Orchis palmata maculata. Differt a praecedente Bracteis minoribus, Labio crenato trifido: intermedia longe minore. Alae seu Petala exteriora patentia, nec, ut in illa, retroflexa. Radices magis patentes & divaricatae ac digitatae.*

Orchis palmata palustris non maculata [*O. incarnata* L.] was different from the *maculata,* since the two outer petals are bent forwards against each other, while in the *maculata* they are spread out. The lip is not as deeply notched and the side lobes are less reflexed, the stalk is shorter, the spike more blunt and the bracts longer than the flowers. It is more similar to *Orchis odore Sambuci* [*O. sambucina*], but the spike is denser and the roots more spread out.

Carex spica simplici androgyna [probably *C. pulicaris* L.] was in flower with a finger-long triangular stalk. There were male and female flowers in the same spike, with the males in the top.

Walnut trees of a considerable size had formerly grown in the garden but had been destroyed last winter.

The gamekeeper told us that the red deer have their heat period in September, the fallow deer in October and that both are pregnant for 40 weeks; that the wild boar mate in October and are pregnant for 16 weeks, like the bear; that there are quite often herons of the smaller type here and large numbers of cranes. He also gave us a few seabirds for description.

The male of "Knipa" (*Anas Clangula*) [*Bucephala clangula* (L.), goldeneye] is white except the bill, back, tail and pinions and a line on the wing which are black and the head which is bluish black; the temples are white and the bill is blunt.

The body of the female "Skräcka" [*Mergus merganser* L., goosander] was grey on the upper side including the tailfeathers and the pinions; but the underside is all white; the head is yellowish grey with a hanging crest.

Cimex hyoscyamoides Petiverii [*Therapha hyoscyami* (L.)] with a red St Andrew's cross on its back, was found in very large numbers on the henbane. We also saw a rather big white spider with a red belly.

Ashwood logs of the size of ordinary timber, 3 fathoms [18 feet, 5.4 m] long and of an age varying between 50 and 60 years to judge from the annual rings lay in the farmyard; they were intended for brancards [horse-litters] at the Court, for which there is no lighter or more suitable wood.

The saltpetre makers at Rälla were in the middle of their boiling; they tested the earth for saltpetre by dissolving it in water; a drop of the water was put on the blade of a knife and dried in the sun, which makes the saltpetre crystallize and turn white. Here the whole process is thus: the earth is leached, the water is poured off and boiled, and diluted with more brine; the boiling is continued for five or six days; the liquid is scooped up into a cauldron, wood ashes are added to remove the scum, decanted, boiled and skimmed again; this is continued *ad pelliculam usque,* or until a drop put on the knife crystallizes immediately. We asked them if a well-leached saltpetre earth will yield saltpetre again after five or six years, and more than untouched earth; this was confirmed. They told us that the best earth is taken from the floors in the cowsheds, but not where the urine falls.

A dry stone wall, as high as a man, some ¼ mile long began here on the left-hand side and enclosed the Crown estate of Halltorp. All along the wall grew *Asclepias* [*Vincetoxicum hirundinaria* Medicus].

Now one could see the nature and peculiarities of the *alvar*-land, which occupies the greater part of Öland; it is a low table-land, all dry, bare and sterile; the bedrock is a red limestone which is partly covered with earth a finger deep, partly bare. From the *alvar*-land to the shore is about ⅛ mile, which is a lowland area adorned with deciduous woods, meadows and tilled land; the slope between the *alvar*-land and the arable land is quite steep, sometimes a perpendicular and very high precipice. Most of the farms are situated at the edge of the *alvar*-land, beneath the slope. The *alvar*-land itself lies like a waste and is useless for the cattle.

The *lantborg* [formerly spelled *landtborg*] is nothing other than the sides of the *alvar*-land, which ends in a little ridge, enclosing the *alvar* in an oval ring especially on the western side. The *lantborg* serves as a main road for travellers, and it needs neither repair nor staking.

Astragalus scapis radicalibus, calycibus leguminibusque villosis, foliolis acutis [*A.* 593 *oelandicus* of Index = *Oxytropis campestris* (L.) DC.], which I have called *Astragalus campestris minimus* in the *Handlingar* of the Royal Academy of Science, 1741: 202, is a little but very rare plant. We first found it on the *lantborg* and on the *alvar*-land opposite Räpplinge church and later in some other places on the *alvar*-land. The root is very long and thin. The plant has no branches but a few creeping shoots ½ a finger long with leaves which are all less than a finger long, all pinnate and consisting of 14-15 paired leaflets which are oblong-oval (*lanceolato-ovata*), glossy with fine hairs. The stem coming from the root is simple and unbranched, bare and as long as a finger, ending in a loose spike with 10-12 flowers, of which the bracts are lanceolate and shorter than the oblong and pubescent calyx, which has five dark points without being grooved. The petals are pale yellow or yellowish-white without streaks. The stamens are diadelphous. Botanists have previously found this flower in the Swiss mountains, and Mr Haller has figured it among his Swiss plants, *Tab. 13.* [Haller, *Enumeratio methodica Stirpium Helvetiae,* 567, t.13; 1742].

Quarries began to be seen in the *alvar*-land. The red limestone, of which the whole *alvar* consists, is quarried for flag stones by the farmers, but it is remarkable that the surface rock is used, never deeper than 6 quarters [3 feet, 90 cm]. The stone is layered in horizontal slabs and the upper ones are softer, thinner and more fragile. As soon as the worker reaches this depth he leaves the spot and does not continue there, hence the district is converted for as far as one can see into an unhappy and stony Arabia, a *terra mortua* or a heap of stone chips. A man can hardly walk on it and nothing will grow here. The fractures are mostly perpendicular but there is an important one in the northwest-southeast direction which the quarry-men follow closely. The stone was quarried with hammer and chisel and dressed into quadratic 18 inch flagstones, and eventually taken to the polishing works, which are mostly in the neighbourhood.

The polishing mills are in the open directly on the ground. The flagstones are laid on the ground in a ring of about 6 fathoms [11 m] diameter. The stone surface is thus slightly elevated above the ground and must be perfectly horizontal. At the centre of the stone ring there is an upright wooden axle and attached to this a *radius,* a 3 fathoms [5.5 m] long bar, with a hole bored in one end where it is attached to the axle. At the outer end is fastened a curved piece of wood, which is almost semicircular and has an iron dowel at each end. The ring is laid with polished stones. Two unpolished stones are laid on the ring and fastened with the iron dowels. A pair of horses or oxen turn the bar around the axle, so that the two stones are polished and also the bottom stones receive their final polish. The stone ring is often sprinkled with water and fine sand or *arena riparia.* The farmers say that they get two *stiver* for a stone, but in Stockholm the price is much higher.

Lepidium foliis pinnatis integerrimis, petalis calice minoribus [*L.* 535 *minimum* of Index = *Hornungia petraea* (L.) Rchb.] or *Cardamine pusilla saxatilis montana discoides, Column: ecphr.1. pag. 273* is a little plant, not hitherto seen in Sweden, which grew in the quarry-waste. *Radix annua. Caulis digiti longitudine atque* [*absque*] *ramis. Folia pinnatifida tribus paribus, extimo impari. Racemus drabae. Calyx tetraphyllus, concavus. Petala 4, linearia, patentissima, modice recurva, apice bifida, altero denticulo minore. Stamina 6, albida, subaequalia, patula, adscendentia. Antherae subrotundae, luteae. Germen compressum. Stylus nullus. Stigma capitatum. Silicula ovalis, horizontalis, depressa, subtus gibba, vix emarginata, dissepimento opposito, nec parallelo.*

Chives [*Allium schoenoprasum* L.], which is often grown in gardens and used for food for men and chickens, is a plant concerning which the botanists do not know where it grows wild. Here among the gravel grew a kind of leek which subsequently we found all over the *alvar*-land, and which is called "alvar-lök" by the Ölanders [*A. schoenoprasum* var. *alvarense* Hylander, 1945], and it is a *Cepa scapis foliisque subulatis teretibus aequalibus, spathis globosis* [*A. schoenoprasum* L.]. I planted it in the university garden in Uppsala, where it is so like the ordinary chives that no difference could be noted, except that the garden chives, when it first shoots up, has the leaf-tips bent backwards but they are not so in this.*

Besides *Lepidium* [*Hornungia petraea* (L.) Rchb.] and the chives there were no remarkable plants in the quarries except *Valeriana locusta dicta* [*V.* 33 *locusta* of Index = *Valerianella locusta* (L.) Betke] or cornsalad. *Herniaria* [*H. glabra* L.], *Abrotana campestre* [*Artemisia campestris* L.] and *Saxifraga tridactylites* [*S. tridactylites* L.] grew along the *lantborg*.

At Borgholm Castle we had the *lantborg* to our right, and the district was full of juniper bushes. We left Borgholm Castle on our left and its outhouses on the right. We passed the inn with its market-place and headed for Köping church.

A cairn of considerable size was seen on our left close to the seashore between the inn and Köping, where a bay cuts ⅛ of a mile into the land; this was constructed of greystones, which are not found in this place, and each stone was so big that it could hardly be lifted; from this as well as from my experience of the Gotland cairns I judge it to be a *tumulus sepulchralis.*

A lichen, called *Lichenoides tinctorium atrum foliis minimis crispis Dill. musc. 188 t. 24 f.81* [*Parmelia stygia* (L.) Ach.] and very rare in Sweden, grew on the north side of the cairn; this moss is pitch-black and could presumably be used for dyeing since it colours paper bright red, but it is not common enough to be of any great use.

Anthyllis flore coccineo [*A. vulneraria* L. var. *coccinea* L.], which Dillenius in the *Hortus Elthamensis,* 431, t. 320 [1732] calls *vulneraria supina flore coccineo* and which he considers to be a different species from the yellow kind, grew beside the cairn together with the yellow one.

---

* *Note.* This observation by Linnaeus was confirmed by Turesson's transplant experiments, recorded in *Hereditas, 6*: 156-159 (1925), which led Hylander in *Uppsala Univ. Årsskr., 1945,* no. 7: 113 (1945) to name the Öland alvar-ecotype, with erect very glaucous relatively short leaves and earlier flowering, *Allium schoenoprasum* var. *alvarense.* —*W. T. S.*

On our right hand we had the *lantborg* on top of a steep cliff 3 fathoms [5.5 m] high, with several vaulted caves in it; in the evening we reached Köping church.

There were several burial mounds around the church, some of them surrounded by flat stones, others marked out with a ring of upright stones. Between the church and the vicarage stood the biggest runic stone I ever saw, which had text all over one side of it, but it was difficult to read because of the crust of *Lichene leproso albo* [*Lecanora calcarea* (L.) Sommerf.] which covered it. The carved serpents were evident, and we were able to read the following letters:

ᚦᚢᛁᛁᛦ·ᛁᛅᚢ····ᛋᛏᛅᛁᚾ·ᛅᚢᚴᚦᚢᚱᚠᛅᛋᛏ······
ᛒᚱᚢᚦᚱ:ᚢᛅᛁᛋᛏ•ᚢᛁᛋᛏᛅᛁᚾ·ᚴᚢᚾᚠᚢᛋᛁ ········

Linnaeus: *þuiiʀ iau . . . stain . aukþurfast . . .*
bruþr : uaistu=u–stain . kunfusi . . .

Öl 46: *þuriʀ auk þurstain : auk : þurfastr : þaiʀ : bryþr : raistu . stain :*
at : kunfus : faþur : sin : kuþ : hialbi : siul : hans

"Tore and Torsten and Torfast, those brothers raised the stone in memory of Gunnfuss, their father. My God help his soul."

In the church was a font made of coarse, porous, almost mushy limestone not unlike filtering stone.

True violets [*Viola odorata* L.], which Dr Linderstolp described as growing abundantly here in Köping, and for which we had made a whole day's detour, we searched for eagerly, but neither here nor anywhere else in Öland did we find any true violets, only three species of dog-violets.

Gooseberries grow wild here, forming very big bushes between the church and the vicarage, and also *Scandix seminibus hispidis* [*S.* 242 *echinata* of Index = *Anthriscus caucalis* Bieb.] and *Chaerophyllum caule maculato, geniculis tumidis* [*C.* 244 *geniculis* of Index = *C. temulentum* L.].

Rooks [*Corvus frugilegus* L.], a kind of crow, had built many nests, like those of magpies, in the trees in Klinta village near Köping. These birds were said to do much damage to the crops and the peas, so that in some places it was necessary to hunt them; this is done by boys who climb the trees where the birds usually rest during the night; this makes the birds move to other trees, but later, when it is dark, they are chased away and, when they subsequently take shelter in their usual trees, the boys catch them.

We stayed tonight in Köping.

June 3 [p. 55]

In the morning we went out to see a pit in Köping in the middle of a field, in the third corner of an equilateral triangle formed by the church and the vicarage; it was not deeper than the height of a man and no rock was seen in it; this is said to have been started by the Rev. Nils Wallin, who hoped to derive silver from it, but it has not been mined for a long time. We found a few rocks which we crushed and there were streaks of arsenopyrite in them.

In a meadow called Kongsängen we found *Behen album* [*Silene vulgaris* (Moench) Garcke], *Tanacetum* [*Chrysanthemum vulgare* (L.) Bernh.], *Convulvulus minor* [*C. arvensis* L.], *Delphinium* [*D. consolida* L.], *Melampyrum spica quadrata* [*M. cristatum* L.], *Statice* [*Armeria maritima* (Miller) Willd.], *Androsace* [*A. septentrionalis* L.], *Sium* [*S. latifolium* L.], *Iris palustris* [*I. pseudacorus* L.], *Thalictrum 2* [*Th. flavum* L.], *Lathyrus clymenum* [*L. palustris* L.], *Scabiosa Morsus diaboli* [*Succisa pratensis* Moench], *Frangula* [*Frangula alnus* Miller], *Mercurialis* [*M. perennis* L.], *Laserpitium* [*L. latifolium* L.], *Ulmaria* [*Filipendula ulmaria* (L.) Maxim.].

Aster salicis folio [*Inula salicifolia* L.] was so abundant that it covered all the meadow.

Orchis militaris hiante cucullo [*O. militaris* L.] grew so abundantly as to be disdained. There was quite a lot of *Orchis minima militaris* [*O. ustulata* L.] and *Orchis rubra longis calcaribus* [*Gymnadenia conopsea* (L.) R. Br.] was common.

Veronica humilis erecta montana, flore parvo caeruleo Dill. App. 38 [*V.* 20 *Dillenii* of Index = *V. verna* L.] grew abundantly on the impoverished slopes.

Änglöken (*Porrum 2*) [*Cepa* 265 *vulgaris* of Index = *Allium oleraceum* L.] grew everywhere. *Radix rotunda; caulis pedalis. Folia subcylindracea, fistulosa, striis septem prominentibus rigidis subtus notata; spatha diphylla foliacea longissima.*

Carex 4 [*C. arenaria* L.] has a composite spike, consisting of many spikelets, of which the top spike has male flowers below the females, but at the base there are several spikelets with female flowers only.

We returned in the direction of Borgholm on another road, inside the *lantborg*. In a meadow opposite the inn we found an *Ossea* [*Cornus sanguinea* L.] or a true "bone-wood", a shrub which I have seen only in Skåne, and *Astragalus 1* [*A. glycyphyllus* L.], which has also been seen in Skåne and on the Hyke mountain in Dalarna.

We walked up to the castle where we searched in vain for violets, but found *Adoxa* [*A. moschatellina* L.], *Veronica hederulae folio* [*V.* 18 *hederaefolia* of Index = *V. hederifolia* L.] and *Prenanthes* [*Lactuca muralis* L. = *Mycelis muralis* (L.) Dumort.].

Borgholm Castle is built on a promontory of *alvar*-land in the northwest side of Öland. The castle is of a considerable size, rectangular, surrounded by walls. The ground between the castle and its walls is filled up with earth to the height of the wall. There were four round towers, one in each corner. The roof is made of wood, but the southern part has neither roofs nor windows. The walls are made of limestone or Öland-stone, unharmed by moisture in spite of their being unprotected for so many years. The castle was said to have been built for King Charles X during his minority and at present had 5 or 6 men as a garrison.

Here we left the main road for Helvetesgrind (Hell's Gate), still looking unsuccessfully for Linder's violets [*Viola odorata*], but instead we found the *Sanicula* [*S. europaea* L.] of the pharmacies abundant in the wood beneath the slope.

Hazel bushes, very small and hardly 1-2 ells [2-4 feet, 60-120 cm] tall but nevertheless with much fruit, grew here at the edge of the *alvar*-land. These small hazel bushes would be excellent for low hedges in gardens, if they would only stay as short as this; however one noticed that the further down the slope we went, the taller the hazel was, and in the lowland it was like ordinary hazel. This means that its shortness is due to the bedrock and not to its own nature.

We now headed for the Halltorp, the Crown estate across the *alvar*-land, where we saw *Astragalus campestris minimus* [*Oxytropis campestris* (L.) DC.] and much *Androsace* [*A. septentrionalis* L.] and a stud with many mares.

We had to stop at Halltorp because of a thunderstorm.

Cochlearia, Coronopus Ruelli dicta [*Cochlearia* 539 *Coronopus* = *Coronopus squamatus* (Forsk.) Aschers.] grew copiously on the farmyard. In the garden, lavender, balm (*Melissa*), rue and crisped mint [*Mentha crispa* L. = *M. spicata* L. "Crispata"] were so luxurious, that one never saw any bigger. To cultivate medicinal plants would be more profitable here than in any other place in Sweden.

Hawthorn grew in the meadow at Halltorp as well as at the polishing mill opposite, 3 fathoms [18 feet, 5.5 m] high and so stout that we could hardly circle the trunks with our arms.

A white clay is found in Högstrums parish near Väster-Sörby in a bog called Västerkärr; the inhabitants used it to whitewash walls and chimneys; at this time of the year the whole bog was under water, but in August it dries up; possibly it is a lime earth. It is said to lose its colour unless mixed with glue.

We saw a little heather on the hills just before we reached the inn, but it was so short that we hardly noticed it.

A yellow earth, which an old woman used instead of buff colour to dye leather, occurred at Källartorpet near Isgärde. It was found close to a well along the road in a little pit on the left side, and the pure yellow ochre the woman used was altogether similar to the one seen in mineral springs. The hides coloured with this dye had exactly the same colour as those dyed with foreign ochre.

Dye-yielding plants used by the inhabitants of the district are "King" or wild marjoram (*Origanum*) [*O. vulgare* L.] for red, bur-marigold (*Bidens*) [*B. tripartita* L.] for orange, and buckthorn or *Rhamnus catharticus* berries for green and bark for brown and yellow.

The alvar-stone used for flag-stones explodes and sparkles when put into a fire so intensely that it might well damage bystanders. We were told that it is used for stoves and fireplaces but they have to be constructed with great care so that the edges, but not the sides, of the stones were turned towards the hearth in the same way as the miners arrange the stones in their blast furnaces.

Lime can be burnt from this stone but it is grey and coarse and can be used for mortar only. It is however seldom used; instead lime is burnt from loose limestone boulders.

We were told that the farmers in this parish used to work in the quarries, and when they did, they were as poor as all quarry-men are but now they have discontinued the quarrying and concentrated on farms and fields and they are much better off.

Hops grow well here, when they are planted, which is seldom in this land, and this is convincingly demonstrated at the vicarage.

"Källerhals" [*Daphne mezereum*] grew in the bogs not far from the church.

Late in the evening we went out to listen to the song of birds called "kledra" here; they turned out to be nightingales, which delighted us with their lovely singing.

We spent the night at Glömminge.

June 4 [p. 59]

In the morning after a night's rest in Glömminge we went out into a field where we found alum slate and several petrifactions, like the "shells" of small insects. There were also cobble stones which were hard on the outside but soft inside so that they changed into sand between the fingers when crushed.

To keep moths away from wrapped up clothes the farmers use woodruff (*Asperula 1*) [*Galium odoratum* (L.) Scop.], wood millet (*Milium 1*) [*Milium effusum* L.], or *Semina meliloti* [seed of *Melilotus altissima* Thuill.] (which grows wild on southern Öland) between the garments. It is strange that odorous things in nature, e.g. amber, musk, civet, abelmosk, woodruff, melilot and wood millet, should be used to drive away *Acari* in clothes as well as in ointments for scabies and in remedies taken for exanthema, scabies and contagious diseases.

Marsh rosemary (*Ledum*) [*L. palustre* L.] is used in the pigstyes when the swine have many lice.

The characteristic flowers here were *Crepis* or *Hieracium majus* [*Crepis tectorum* L.], *Lapsana* [*L. communis* L.], *Ballota* [*B. nigra* L.], *Chaerophyllum geniculis tumentibus* [*C. temulentum* L.], *Cynosurus 1* [*C. cristatus* L.] and *Avena spicis erectis* [*A. pratensis* L.].

Ålandsrot *(Enula)* [*Inula helenium* L.] grew here very sparsely.

Bryonia [*Bryonia alba* L.] grew here and there along the stone walls. The local name for the *alvar*-leek [*Allium schoenoprasum* L.] is hundlök, [houndsleek], for *Ulmaria* [*Filipendula ulmaria* (L.) Maxim.] tulk-grass.

The kind of chives which the Gotlanders call "keipe", or *Porrum, capitulo bulboso erecto, foliis planis subcrenatis, vaginis ancipitibus* [*P.* 266 *Kaipe* of Index = *Allium scorodoprasum* L.], is here used like cabbage. Neither I nor any other botanist has seen this leek in Sweden before. When it is young it looks altogether like a garden leek, but when it grows older it looks like a sand leek. *Radix subrotunda oblonga. Caulis pedalis vel bipedalis, teres, vaginae foliorum compressae, utrinque carinatae; carina altera per dorsum folii excurrente, altera vero sinu folii terminata; faux folii instruebatur membranula tenui caulem arcte amplectente. Folia plana solida, spithamea, digiti transversi latitudine, margine cartilagineo-serrata. Spatha diphylla acuminata.*

The roots of *Saxifraga alba officinalis* [*S. granulata* L.] are dried, pulverized and taken by the farmers for pleurisy, to what good I cannot see. *Cynoglossum* [*C. officinale* L.] had white flowers and all day we saw chicory [*Cichorium intybus* L.] growing wild.

There was a beacon at the roadside not far from Glömminge. Its construction was quite strange, a tall perpendicular mast on a tripod with room for a man under it. Juniper bushes were hung on it, like sheaves, on hooks all the way from the feet to the top of the mast.

The stones in the old fences were all covered with a white *Lichen leproso crustaceo* [*Lecanora calcarea* (L.) Sommerf.].

Algrum church, which is both an antiquity and a landmark for seafarers, was seen on the left and opposite it were a few burial mounds.

Björnhovda meadow, which gave us so many beautiful flowers before, now showed us *Scrophularia* [*S. nodosa* L.], *Carduus palustris* [*Cirsium palustre* (L.) Scop.] and *Ophrys major* [*Listera ovata* (L.) R. Br.], which had a honey pouch in the lip, from the base to the notch; its stamens were slightly different from those of *Orchis*.

The meadows were in flower with the little *Primula* [*P. farinosa* L.], *Ranunculus acris, Pinguicula* [*P. vulgaris* L.], *Scorzonera* [*S. humilis* L.], *Saxifraga* [*S. granulata* L.], *Acetosa* [*Rumex acetosa* L.]. In the groves there were *Orobus vernus* [*Lathyrus vernus* (L.) Bernh.], *Anemone nemorosa, Geranium aconitifolium* [*G.* 572 *aconitifolium* of Index = *G. sylvaticum* L.], *Melampyrum* [*M. pratense* L.] and *Alsine foliis lanceolatis serrulatis* [*Stellaria holostea* L.]. *Beccabunga* [*Veronica beccabunga* L.] grew in the marshes, elm trees and crab apples around the meadows.

Limax cinereus maculatus [*L. maximus* L.], a big wood slug, was seen under the trees. It was black with a furrowed back and wavy fragmented ridges, the breast looked like shagreen, it had four small horns and the pore was on the right side of the breast.

All the way from Björnhovda to Torslunda the road passed through delightful groves.

Insect nests [of *Phalaena neustria* (L.), according to Schreber], as big as small heads, stuck on the blackthorn, consisted of a white tissue with many vaulted chambers and labyrinths inside. The leaves were eaten up. The grub (*eruca*) was black, with 16 feet. On every segment of the body there was a yellow ring on each side with brown short hairs within it. The whole body was covered with long white hairs. Some grubs were black and had brown spots within the circles, others were all white with red-brown middle and hind feet. The threads were tougher than silk, and could possibly be utilized.

We hurried through the meadows in Torslunda. Here we saw *Statice* [*Armeria maritima* (Miller) Willd.], *Alchemilla 1* [*A. vulgaris* L.], *Actea* [*Actaea spicata* L.], *Veronica scutellata* [*V. scutellata* L.], *Sanicula* [*S. europaea* L.], *Stachys foetida* [*S. sylvatica* L.], much *Ophioglossum* [*O. vulgatum* L.], much *Ophrys* [*Listera ovata* (L.) R. Br.] and still more *Paris* [*P. quadrifolia* L.].

Leontodon 1 or *Taraxacum* [*T. officinale* Weber] grew in a marshy meadow with entire leaves, without the slightest tooth, *lineari-lanceolata.*

Papilio albus, subtus viridi colore marmoreatus, Petiv. mus. n. 306 [*Anthocaris cardamines* (L.)] and *Tenebrio 1* [*Blaps mortisaga* (L.)] were found among the bushes. The latter had antennae with eight joints and pointed claws at the mouth without any teeth.

Sheep-tick [*Ixodes reduvius* (L.)], a species of *Ricinus* or *Acarus,* which damages the wool of the sheep very much, here as well as abroad, was seen today. It is similar to a bedbug in colour and size, each leg had five or six joints. The fore feet are slightly bigger than the rest. The belly is compressed and flat and it lacks spots on the back, and an elevated ring encircles the whole body.

The beehives suffered much damage last year and there were not many bees left.

In this parish are many destitute people who support themselves collecting bast, acorns and nuts; there is more oak, hazel and linden here in western Öland than in any other district in Sweden.

The bast is taken from young linden trees, from last year's shoots, without retting.

The acorns are sold to Kalmar for 8 *stiver* a bushel and used for fattening swine.

Nuts are sold for 10-16 *stiver* a bushel and taking nuts from the trees before St Bartholomew's Day [24th August] is strictly prohibited.

"Mouse nuts" are the best nuts, collected by mice, who store them in their nests under hazel bushes and tussocks as winter provender, for one can often find almost a bushel in a nest without a single nut being worm-eaten or hollow.

When we left Torslunda we left the main road towards the right in order to ride on the coast road beneath the slope. This was the first time we saw the Öland "tok", a plant which we had been very eager to find, since several members of Parliament had mentioned it to us.

The "tok", as it is called on Öland, is a bush which is extremely rare in the world, for hitherto botanists have seen it only in York in England, and recently in Siberia, and now in south Öland. This bush is described by Ray, Morison, Miller, Walther and Amman, but by none of them well described. The Latin name is *Potentilla caule fruticoso* [*P.* 416 *fruticosa* of Index = *P. fruticosa* L.]. It grew in tussocks in the *alvar*-land, beside low places where water stays the whole winter. It is as big as lavender or hyssop, has yellow flowers and sheds the outer bark layer every year. This bush is used for low hedges in gardens, where it grows well. The Ölanders make no use of it except that they make whisks for scouring kettles from the hard and rigid twigs.

The alvar-leek [*Allium schoenoprasum* var. *alvarense* Hylander] grew frequently together with the tok, but not in tussocks like chives.

As we walked over the *alvar*-land towards the slope at Eriksö we saw a slide in the cliff where the slate was bare. When tested with the tongue this slate had the taste of alum. There was also one slate with green and red grains, like verdegris with green patches, a reliable indication of copper content. Between the slate layers was a fine alum-containing clay.

We travelled towards Resmo along the bottom of the slope, which was very steep with bare cliffs on our left, and the sea on our right. The road passed through the most beautiful groves one could ever see, which for beauty surpassed all other places in Sweden and competed with all in Europe; they consisted of linden, hazel and oak, with the ground smooth and green, without rocks or moss; here and there were the most delightful meadows and cornfields. A more pleasant resort cannot be found for anyone who has grown tired of the capriciousness of the world and wishes to retire from its vanity into a quiet obscurity.

The ducklings were big as recently hatched goslings, dark on the head, neck, back and wings and white elsewhere with a black bill. They were running about in the groves and could be caught with the hands.

One saw deer in herds here and there, fat and shiny with white spots.

Milium floribus dispersis [*M.* 55 *odoratum* of Index = *M. effusum* L.], a tall fragrant grass, was frequent in the groves, and *Dentaria* [*D. bulbifera*], *Mercurialis* [*M. perennis* L.] and *Alliaria* [*A. petiolata* (Bieb.) Cavara & Grande] grew abundantly. A variety of the little *Primula* [*P. farinosa* L.] had white flowers.

An eighth of a mile before we came to Resmo church we climbed the cliff at the steepest place, to look for alum where the rock surface was bare.

Cochlea testa supra convexo-plana, subtus convexa perforata, anfractu acuto, apertura semicordata [*Helix albella* L.] was found at the rock-slide. It is remarkable because the shell has a sharp edge.

Globularia caule herbaceo foliis radicalibus tridentatis, caulinis integerrimis [*G.* 109 *montana* of Index = *G. vulgaris* L.] grew on top of the *lantborg.* It is rare in Europe and has not been observed in Sweden before. I have seen it on the hills at Fontainbleau in France, but here in Resmo it had longer roots and bigger flowers.

Radix perennis. Folia radicalia, petiolis longissimis innixa, ovata, tridentata, intermedia productiore, utrinque glabra, nitida, viridia. Caules spithamei erecti, simplicissimi, striati, purpurascentes: Foliis caulinis minimis, alternis, lanceolatis, infimis tridentatis. Flos terminatrix, solitarius. Corolla caerulea, corollulis irregularibus monopetalis, bilabiatis; labio exteriore majore, oblongo, trifido; interiore erecto, minimo setaceo. Staminibus quatuor, Pistillo unico. Paleae flosculos destinguentes acuminatae & fere pungentes.

Cistus Oelandicus Rudbeckii [*Helianthemum oelandicum* (L.) DC.] grew on the *alvar*-land, very unlike the common *Helianthemum*; the leaves were smaller and smoother. The flowers were also smaller, like those of *Potentilla* without a spot inside, the petals so narrow that they did not touch each other at the sides.

We saw several "älvdanser" [fairy dances, i.e. fairy rings] in the meadows beneath the *lantborg.* When we examined them closely we saw that they consisted of *Cynosurus bracteis integris* [*Sesleria caerulea* (L.) Ard.], a grass with blue leaves which grows in a ring. When this grass grows in a shallow soil, it looks like a blue circle, which common people believe to be caused by the fairies' dancing. Scientists have attributed it to the nature of the soil, to exhalations or to horse dung. But here it was quite clear that "fairy dances" are nothing but this grass, which spreads in all directions and finally withers in the centre, creating the ring.

The crops are so good here in Resmo that the inhabitants are able to sell grain and buy wooden vessels for household use from the northern parts, where there is forest.

Firewood has to be bought from Småland, here as in all the southern district.

The fields are not left fallow unless the soil is sandy. Then it rests every 5th or 6th year, but in Glömminge and Torslunda, where the soil is sandy too, crops are grown for three years and the field lies fallow for two years.

Hops are too difficult for the farmers to grow here because there is no wood for poles etc.

Moles are few here. Weasels are numerous and swans are seen only in spring.

"Alle" [*Uria grylle* (L.), black guillemot] is a seabird as big and black as a raven, presumably identical with the "Grylle" of Gotland.

The nightingales were singing beautifully in the evening. We stayed the night in Resmo with the rural dean, Johan Wallin.

June 5 [p. 67]

A few trees were observed so far away from us on the *alvar*-land east of Resmo that one could hardly see them. The inhabitants called them "gran" [spruce]. The botanical student Samuel Wendt was sent to examine them and returned with a twig of *Taxus foemina* [*Taxus baccata* L.] or yew. He said that there were three such trees, each 3 fathoms [5.5 m] tall from root to top, densely covered with branches; one of the trees had four trunks emerging from one root, the original trunk having been cut down.

We saw a few burial mounds, one of which, in a field north-west Resmo church, was remarkable. There was a circle of nine big stones at a distance of 5 ells [3 m] from each other. Each of these stones rested on three smaller stones, and the tenth stone was in the middle.

The stone walls are much stronger, broader and higher here than in other places. All the stones in the wall are laid horizontally so that the cattle cannot trample them down.

A rook [*Corvus frugilegus* L.] had built a nest in a high tree in Resmo. We had it taken down and found that it was a hemisphere one foot in diameter, built of straw, bast, feathers, hair, moss and turf at the bottom. Inside were three nestlings, all black, big as magpies and almost fledged. The bill looked like that of a crow, black, somewhat lighter in the middle. The mouth was broad and the bristles covering the bill were very stiff. Mites (*Culex alis aqueis, pedibus nigris, annulo albo*) [*Simulium reptans* L.] were in great abundance in the nest, all bloated with the blood of the young birds.

An oak trunk, which was very big and cut down last winter, had a diameter of 7¼-ells [3½ feet, 105 cm], the bark excluded. As we counted the annual rings we found it was 260 years old and also noted that some rings in the wood were very close together and others much further apart. As I considered what had caused this, it occurred to me that cold winters might cause closer rings; thus I counted the rings from the bark towards the centre to the years 1708-1709, the year of the very cold winter, for which I found the rings very close to one another, likewise for the years 1587 and 1658. I noticed the same on a great number of other smaller oak logs. Thus we have in the oak trunk a chronicle of winters, from which we can tell the weather 200-300 years back.

A stream, the biggest and most beautiful in the whole island, ran beneath Resmo farm. It was strong enough to drive a little mill, albeit with water shot from above.

There are many windmills on the *lantborg*, especially on the western side.

Myslinge mound stands ⅛ of a mile from Resmo. This is one of the highest mounds in the country. It lies on the *lantborg* and consists of loose stones. From here we saw Mörbylånga, Kastlösa, Kalmar, Möckelby, Torslunda and Resmo to the west and Sandby, Stenås and Hulterstad to the east.

From the top of Myslinge mound we saw the *alvar*-land quite brown, with green lines, its length and breadth laid out like a map. We investigated the cause of this and found that the bedrock, which was covered with a dry sterile brownish soil scarcely a finger deep, was fractured like ice in extreme cold. Grass grew over these fissures where the moisture was retained longer and not so easily dried out as on the flat surface.

From here we rode down the slope to Mörbylånga, an annex of Resmo.

A little fish (*Gasteroesteus aculeis in dorso decem*) [*G. pungitius* L., stickleback] was found in a brook.

Fucus [*F. vesiculosus* L.] or seaweed is used here for covering the huts used by farmers, who have to set watchers to protect the crops from deer and boar during the summer.

The Mörby redoubts here on the seashore had been so corroded by time that little remained even of the ruins. We found a stag-beetle (*Scarabeus Cervus volans dictus*) [*Lucanus cervus* (L.)] on the seashore. The antlers had three big teeth, the antennae had four leaf-like structures at the end and the antennae above the mouth had three joints. The mouth contained two woolly tongues.

From here the road passed through the loveliest groves. In the Årsör meadow *Satyrion* [*Platanthera bifolia* (L.) Rich.] and *Orchis morio* were in great abundance, but *Orchis hiante cucullo* [*O. militaris* L.] only here and there. The oaks on the slopes were still leafless. We do not know if the cold from the marshes or the last winter had ruined them since all the other oaks in the district were in full leaf.

The gamekeeper in Årby gave us an owl [*Strix aluco* L., tawny owl] for description and he also went with us to shoot an "alvargrim" [*Charadrius apricarius* L., golden plover].

Noctua magnitudine erat gallinae, Color griseus; in dorso, collo, capite lineolae fuscae longitudinales, praeter minores undulatas. Venter albus, lineis ut in dorso. Pedes alba lanugine punctis nigris. Oculorum orbita cingebatur circulo griseo, nigro-maculato. Iuxta aures erat circuli segmentum nigrum. Plumae intra circulum erant simpliciter pinnatae raro supra decompositae.

It was noteworthy that the *margo exterior remigum anteriorum* was so serrated that the tips of the threads on the pinna (on the side near the tip of the same pinion) were covered with yet smaller threads.

We now came to travel through Borrum, a very remarkable place. A big grove of hazel and oak lies there beneath the slope (*lantborg*), with a field in the middle 45 paces long and 57 paces broad. On this beautiful field the inhabitants have gathered since time immemorial on Whit Sunday and Whit Monday, to play and dance after the service. The clergymen are much against this ancient tradition, protesting that many a girl has danced herself to ruin here.

The very old Bärby redoubt is situated on the very edge of the *lantborg*, and we climbed the slope to see it. It is surrounded by a semicircular wall with a radius of 90 paces, which ends at a precipitous cliff. On the steepest side there grew in abundance *Lithospermum* [*L. officinale* L.], of which up to now I have only seen a few plants in the island Kofsan in the lake Mälaren and of which the seeds are used in pharmacies. On the north side of the wall a beautiful *Hedera* [*H. helix* L.] grew on the rock. Still further north there was *Cotoneaster* [*C. integerrimus* Medicus] in flower and *Ranunculus foliis ternatis integerrimis* [*R. illyricus* L.]. See June 2, Hulterstad.

A big stone tumulus stood not far away on the summit and had the same *Ranunculus foliis ternatis* [*R. illyricus*], and beneath the summit grew a magnificent *Hedera*.

We saw innumerable burial mounds on the slope all day, covered with earth and surrounded by stones. Nearer Kastlösa there were many upright stone slabs, among which the most remarkable were two narrow slabs, 4-5 ells [2.5-3 m] high at a few fathoms' distance from each other.

Tok [*Potentilla fruticosa* L.] grew on the summit of the *lantborg* and on the *alvar*.

Anthyllis flore coccineo [*A. vulneraria* L. var. *coccinea* L.], which we found on the 2nd of June on the mound at Borgholm, abounded on the *alvar*, where it grew wherever there was a patch of dry brown earth. The belief of gardeners, i.e. that a coloured earth around the roots will yield coloured flowers, might be confirmed in an experiment with this brown earth.

Cistus Oelandicus Rudbeckii [*Helianthemum celandicum* (L.) DC.] was today common on the *alvar* and differs from *Helianthemum vulgare* in that: the flowers are much smaller, the sepals not bent back, the ovary or *germen* is smooth, the style curved, the stigma rough, four-parted and the stamens shorter with the anthers light yellow, not dark gold. The petals are narrower and notched (*emarginata*). The stalk or *scapus* is red, straighter and smoother. The leaves are more pointed and their margins are not revolute. The flowers are most often closed. With so many differences it is evident that this is distinct from the common *Helianthemum* and is a plant which in Sweden has only been seen on Öland and is rare in all Europe.

The following plants grew on the *lantborg*: *Globularia* [*G. vulgaris* L.], *Cistus Oelandicus* [*Helianthemum oelandicum* (L.) DC.], *Astragalus campestris minimus* [*Oxytropis campestris* (L.) DC.], *Lepidium foliis pinnatis* [*Hornungia petraea* (L.) Rchb.], *Rubia cynanchica* [*Galium triandrum* Hylander], *Androsace* [*A. septentrionalis* L.], *Silene quae Muscipula montana hirsuta Ruppii* [*S.* 366 *muscipula* of Index = *S. nutans* L.], *Sedum acre* [*S. acre* L.], *Sedum minus teretifolium album* [*S. album* L.] with white as well as reddish leaves, *Valeriana Locusta dicta* [*Valerianella locusta* (L.) Betke], *Absinthium vulgare* [*Artemisia absinthium* L.], *Saxifraga tridactylites* [*S. tridactylites* L.], *Ranunculus radice globosa* [*R. bulbosus* L.], *Abrotanum campestre* [*Artemisia campestris* L.], *Leontodon* [*Taraxacum officinale* Weber], *Quinquefolium folio argenteo* [*Potentilla* 417 *alba* of Index = *P. argentea* L.], *Quinquefolium minus repens luteum* [*Potentilla* 419 *adscendens* of Index = *P. verna* L. = *P. crantzii* (Crantz) G. Beck], *Geranium lucidum* [*G. lucidum* L.].

The steep part of the *lantborg* had its own plants, such as: *Veronica spicata minor* [*V. spicata* L.], *Globularia* [*G. vulgaris* L.], *Rhamnus catharticus* [*Rh. catharticus* L.] and *Virga sanguinea* [*Cornus sanguinea* L.].

The plants in the *alvar*-land deserve attention, since they thrive in the driest and most sterile soil: *Galium luteum* [*G. verum* L.], *Pilosella uniflora* [*Hieracium pilosella* L.], *Carlina annua* [*C. vulgaris* L.], *Quinquefolium minus repens luteum* [see above], *Rubia cynanchica corollis trifidis* [*Galium triandrum* Hylander], *Sedum acre* [*S. acre* L.], *Sedum petraeum* [*S. album* L.], *Anthyllis* [*A. vulneraria* L.], *Herniaria* [*H. glabra* L.], *Helianthemum vulgare* [*H. chamaecistus* Miller], *Cistus Oelandicus* [*Helianthemum oelandicum* (L.) DC.] and *Festuca* [*F. ovina* L.], called "fårgräs".

An eighth of a mile from Kastlösa church the district between the *alvar* and the sea was all bare. Trees had vanished and we saw no more of the delightful groves which had pleased us so much. To the left and on the *alvar*-land was a village with woods and fields, extending to the east side of the island.

"Alvargrim," a bird said to live only on the *alvar*, and whose name we had not heard of before, awakened in us a keen longing to see one. The gamekeeper, whom we sent out, caught up with us on the *alvar* with a shot "alvargrim". As soon as I saw it I realized that it was very similar to a bird which I have seen before only in the mountains of Lapland, where it breeds among other species of its family and which the Lapps call "houti" [*Charadrius apricarius* L., golden plover]. This *Pluvialis* is as big as a dove. Head, neck, back, wings and tail are mottled in white, black and yellow-brown. This is because each feather is black with a few yellowish brown spots on the sides and lighter spots at the tip. The down at the base of the feathers on the sides of the neck towards the breast is light grey. The underside of wings and tail are white, and the entire breast and the central part of the neck are black. The bill is black and not longer than the head, with a bulge at the end. The legs are grey and the thighs half naked. The innermost joint of the outer toe is joined to the middle toe. The pinions are dark, and there are eight outer pinions, the outer being the longest and the remainder tapering inwards, all being pointed at the tip and white in the middle. The rest of the pinions are also dark but blunt, equally long, with white points, and the same applies to the coverts. The tail is not cloven and there are ten tail feathers with eight lateral lines, pale yellow on both sides.

Adonis radice perennis [*A.* 456 *octopetala* of Index = *A. vernalis* L.], a plant which adorns our finest gardens, was seen in the meadow close to Kastlösa church, growing like small bushes. We had never thought we would find it growing wild in Sweden. The root is perennial. The stalk is ½ an ell [30 cm] tall, branched and withers in the winter. The leaves are more intensely green than those on any other plant. A large pale yellow flower with orange anthers sits on each stalk. I do not understand why it should have branches, since not one of them carries a flower, only the main stalk. The botanists believe that the root is the *Helleborus Hippocratis.*

Thlaspi vaccariae folio [*T.* 531 *incanum* of Index = *Lepidium campestre* (L.) R. Br.] grew in the cornfields. Its leaves were wavy and serrated.

Quinsy or *angina* had raged widely on Öland this year; it makes the face swell up to a terrifying size. Only a few have died from it, and those who died had a pleurisy first, which killed them rapidly.

The people here have recently changed their manner of constructing the corners of their log houses. Previously when the logs projected beyond the corner joint they were liable to be destroyed by rotting due to the intense rains brought by N.E. and S.W. winds. This does not happen so easily with the ordinary dove-tail joints.

Necydalis elytris apice lineola alba [*N. minor* L.], a very peculiar insect, was found in this place. It has rather stout thighs which are almost globose at the end, similar to the *Leptura.* The antennae are twice as long as the body, like the *Cerambyx.* The wing-cases are not even half as long as the belly, like *Forficula,* but the wings are not wholly covered by the cases and they cross as on a *Cimex.*

The weather was cloudy in the morning with a strong wind and a little rain, but the afternoon was pleasant.

We stayed in Kastlösa during the night.

June 6 [p. 74]

We had been told that there was coal somewhere in Kastlösa, and after much questioning we were informed that the place was near Dalby inn. In the early morning we went to Dalby, but the innkeeper who had found it was in Småland and his servants were suspicious. They did show us a piece of coal, but it looked as if it were English. The domestic servant showed us a place north-east of the house at the border between the tilled lands and the meadows where he said the innkeeper had found the coal when he was ploughing. Then the daughter showed us a spot in a slope close to a mound, where the earth had not been touched with a tool for ages, and where there was no rock. In other words, they tried to mislead us in every way although we tried to win their confidence by civility, eloquence and gifts. Finally we left them, having been told that the learned and worthy Jacob Faggot, inspector at the Land Surveying Office, had been there before with the innkeeper to look at the coal at the time he started the alum works here.

The roofs are thatched with straw in this district instead of birch-bark. This is done in such a way that the straw is all covered and as fireproof as a birch-bark roof. The straw is cut off at the eaves where it is covered with a board and a little birch-bark. The straw roof is then laid with double sod layers, the under layer with the grass against the straw, the upper layer with the grass surface upwards. Such roofs are used for the dwelling-houses only, not for the outhouses.

All the villages and farms we had seen so far have been built beneath the *lantborg,* with fields, meadows and the sea on one side and the *alvar*-land with its fresh-water sources on the other.

The maidservants wear blouses, belts and light skirts when working indoors. Their hair is plaited with a ribbon or, as they call it themselves, a *tippat.*

From Kastlösa we travelled over the *alvar*-land to Smedby.

Today as well as yesterday on route for Ottenby we passed many barrows on the *lantborg,* most of them surrounded by stones in a circle, but sometimes the stones were placed in a triangle. There were upright stones both large and small everywhere; but neither in the interior of the *alvar* nor on the shore did we see any signs of graves; thus the ancients have chosen the most spectacular site for a resting place.

The *alvar* looked as before—flat, barren, with tussocks here and there as if it had been rooted by swine. Here and there were pools of water like little lakes but never more than a foot deep, which dry up during summer.

Smedby church was reached ½ a mile from Kastlösa. It was built in such a strange way that I have never seen anything like it; consisting of porch, nave, ante-room and chancel, and the only access to the different parts of the church was by means of low doors, α. The broadest part was the porch, which was

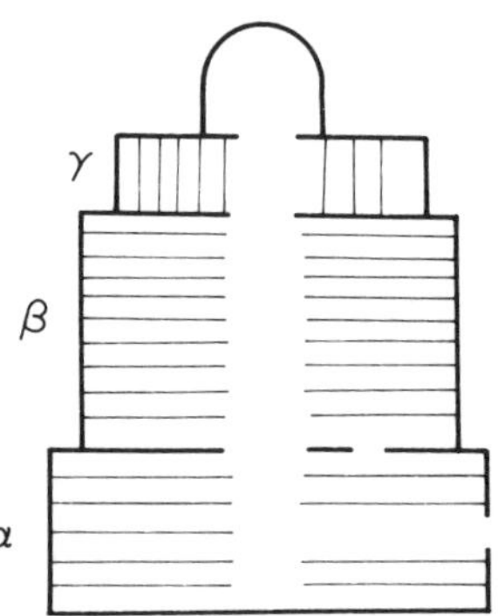

vaulted in the north-south direction with the main entrance in the south wall, β. The nave was narrower but longer and was reached through two doors, one of which was close to the main entrance and none of them higher than a tall man, γ. The anteroom was still smaller and was also reached through a narrow passage. The church walls were a full fathom thick. There were a few windows, each a quarter broad and three quarters high. The church was consequently rather dark, and the congregation might well fall asleep. Few parishioners had the opportunity to see the vicar at the altar, and those who did could not see him when he was preaching, especially those sitting in the porch, since the porch was part of the church. These features indicate that the church is very old. Outside the porch was a grave with a runic stone on which we could read the following letters:

····ᛅᚱ:ᛏᚢᛏᚴᚢ····ᚢᛡᛂᚠᛏᛁᛣ ·ᚴᛁᚾᚢ····

Linnaeus: . . . *an : eftiiketil . o-rsiokuþ*
Öl 12: *aynteis . . . þina : eftiʀ : ketil . boþyr : sin . kuþ . h . . . kus . moþiʀ*

"Östen . . . this (stone) in memory of Kettil, his brother. God and God's mother h(elp his spirit)!"

In Smedby we were told that there was a runic stone in the farmyard in Arvlösa, 1/8 of a mile away. We had already passed the place, but I returned only to find that the stone had been so worn by the ravages of time that only the following letters were legible:

·····ᛏᛏ:ᛂᚠᛏᛁᛁᚴᛂᛏᛁᛚ·ᚭᛂᚱᛋᛁᛡᚴᚢᚦ.

Linnaeus: . . . *nr : auaku . . . uheftiʀ . kinu . . .*
Öl 10: *eimunr : auk : kuna- . . . þair . -uku : kirþy : þisn : eftiʀ : kinu : fatran*

"Emund and Gunnar . . . they raised this memorial after Gunne the bold man."

Our journey took us through the meadows between Smedby and Möckelby, where *Adonis radice perennis* [*A. vernalis* L.] and *Cichorium* [*C.* 650 *agreste* of Index = *C. intybus* L.] grew especially around Klinta. This *Cichorium*, with its beautiful blue flowers, grew here in sufficient amounts to supply all the pharmacies in Sweden, more appropriately for medicine than for food, and has been found only in Skåne before. The ryefields were yellow with charlock and the barley fields were infested with *Serratula Carduus in avena dicta* [*Cirsium arvense* (L.) Scop.].

A runic stone stood on the lefthand side near Möckelby so high that one would have to be on horseback to be able to read anything of it. We copied the following letters, intending to continue the copying after having visited the mine.

ᚴᛂᚠ·ᛁᛁᛣ:ᛋᛏᚢᚱᚴᛅᚱ············

Linnaeus: *kef . iiʀ : sturkar . . .*
Öl 13: . . . *t . . . reisti : stein . -ina × eftiʀ . styrkar . fa . . .*

". . . raised this stone in memory of Styrkar, fa(ther) . . ."

Möckelby lay 2 miles from Kastlösa, where the *lantborg* between the church and the shore runs almost to the edge of the sea and here the alum works are situated.

The alum mine is about a gunshot [about 225 m] in diameter and about 12 ells [320 cm] deep; it forms a semicircular excavation in the *lantborg* on the western side, the bottom being level with the shore and the sides as high as the *lantborg.* The walls are all perpendicular. The ore is here the easiest to work one could ever wish for, since it is nicely partitioned by seams and faults. We counted the layers in the rock surface and found the following: (1) turf with a hand's breadth of mould; (2) slate, one inch; (3) sandy earth, ½ ell [30 cm], the upper surface similar to coal; (4) slate ¼ quarter [3 cm]; (5) anthraconite [bituminous limestone]; (6) blackish slate ½ quarter [about 6 cm]; (7) anthraconite, ½ quarter [about 6 cm] in small pieces; (8) slate one finger; (9) anthraconite with some lime under it; (10) slate mixed with anthraconite; (11) limestone down to the true alum slate 6 ells [360 cm] deep.

"Svartbåller" [Black balls] is the name for a kind of big black stone, shaped as a round loaf of bread. They are so very hard that one can only break them with the greatest difficulty, even more than with flint. The broken surface is pitch-black. When they are put into a fire they make a sharp bang like a pistol shot, and produce a fume with a pungent smell of garlic which gives rise to a sweet taste on the tongue and is rather dangerous, since it contains arsenic. It is a peculiar feature of these stones that, in spite of being harder than flintstone, they are dissolved by rain and sun and disintegrate to mould, and still stranger is the fact that a piece of rock can be turned into a black mould.

Here and there between the beds of slate were some small crystalline grains like rock salt. First one thought them to consist of pure alum, but when one tasted them, they did not dissolve in the mouth. When one looked closer at them they appeared more like a spar, every particle being a rhomboid crystal with two large flat *plana rhomboidea opposita* and on every side two *rhombis oblongis juxta se positis,* thus these crystals were a *selenites.*

One saw no pyrites, but there was said to be plenty of it in the deeper beds, which were avoided because of the water, and there is enough ore in the surface layers.

Lime is burnt with this alum slate as fuel. When pulverized it burns like coal. We asked why lime is not produced by burning the anthraconite, which is likewise found here so plentifully, since this yields an excellent lime, if not a gypsum, but we were told that, when anthraconite burns, it turns into a powder even before it is slaked, and thus it cannot be burnt with the slate, since the two cannot be separated after the burning.

Earth, stone and alum slate had been piled up to form a slope from the bottom of the mine to the shore in order to facilitate the transport of the ore, which is loaded on boats and shipped to Lovers' alum works in Småland. These as well as Kristianopel can be seen from the mine. A few years ago, the stone heap had caught fire, and burning continued on the north side without flames, but smoke and fumes emerged from the ground we walked upon, which was quite warm. When we poked in the ground with a stick, a little smoke emerged, and we found burnt lime, alum, flower of sulphur, red slate and sintered pieces of alum. We asked the inhabitants if they were not afraid that the fire should assail the ore within the mountain, but they thought this was impossible as long as the slate was compressed in the strata. Nevertheless it is my opinion that, if the slate once were to catch fire, Öland might have the first volcano in Sweden.

It was to my misfortune that I was not content to count the beds in the mine from a distance but ordered some of the men to climb the rock wall and throw down a specimen of each stratum; a big boulder fell down and burst against the wall and a fragment hit my left ankle; I was sure that the bone was fractured and my foot crushed. That minute the pain took away all pleasure in the mine or in any further investigations. At noon, I was carried to the inn in order to attend to the contusion.

We excavated a barrow when I was resting at Möckelby inn and found a lot of bones in it, which made us wonder whether people were taller in ancient times than nowadays. We took a femoral bone and inserted it between two boards to be able to measure it accurately. It was 3 quarter-ells and 4½ inches [about 56 cm] according to the measuring rod, thus 4 quarter-ells but for two inches. But the *tibia* was 3 quarter-ells and 1 inch [about 47 cm] ; thus the man whose bones these were must have been almost four ells [about 240 cm or 8 feet] tall.

We continued our journey to Albrunne village; before we arrived there we saw two upright stones 6 ells high and 1 ell broad among the barrows. We passed Albrunne and arrived at a beautiful oak grove where we saw *Laserpitium* [*L. latifolium* L.] , *Hepatica* [*Anemone hepatica* L.] , *Pulmonaria* [*P. officinalis* L.] , *Alliaria* [*A. petiolata* (Bieb.) Cavara & Grande] , *Astragalus* [*A. glycyphyllus* L.] , *Viola martia foliis trachelii* [*V.* 718 *trachelifolia* of Index = *V. hirta* L.] and *Viola martia flores ad radicem, semina in cacumine ferens* [*V.* 719 *monoica* of Index = *V. mirabilis* L.] .

The latter violet had heart-shaped, rounded and slightly pointed leaves; the stalk is rounded but flat on one side and ends in two leaves embracing a triangular calyx, which consists of five sepals, the two inner ones being rather narrow; on the back of the calyx there are three big wavy bracts and two small entire bracts; there were no petals, but stamens and a style; the male flowers at the root with their large corollas were already withered.

Along the road to Vickelby church we saw many large hawthorn trees; a furious storm from the sea raged over the plain in a demonstration of its strength but these tough and strong trees withstood the powerful gale. We learned from this that those who want to plant gardens in woodless and flat districts should first plant hawthorn around them to give protection to the small trees and prevent the big ones from being overthrown by the gales.

A wall, as high as a man on horseback and more than 1 ell broad, laid with horizontal flat stones, stretches across Öland and cuts off a park at a distance of ½ mile from the southern point, wherein is the Crown estate Ottenby. The manor house is in the western part on the plain, but the eastern part is wooded.

Ottenby House is built of wood. We arrived there in the night, but we had to travel further, since the inhabitants were afraid of us, although we explained the purpose of our journey.

Therefore we continued eastwards for ¼ of a mile to Ås church, which stands on the eastern side of Öland. The *lantborg,* which had from Brogholm gone from north to south, stretched from Ottenby to Ås in the east. From Ås church the *lantborg* went along the east coast northwards to Näsby inn.

"Tok" [*Potentilla fruticosa* L.] grew between the *lantborg* and Ottenby estate, tall and abundant; we saw that the oxen ate the leaves and the smallest twigs, which makes the bushes appear uneven, shapeless and gnarled.

We spent this night in the vicarage of Ås, but due to the pain I was unable to get even a minute's rest. I, who so often helped others in similar and other troubles, had not the slightest remedy to help myself.

June 7 [p. 83]

This day was a Sunday, hence we rested here all day.

The reason why this land is delightfully green, and not yellowish or brown like in the rest of Sweden, is that there are no mosses such as *Polytricha, Brya, Hypna* or *Sphagna* except for a few rare plants on the hillocks. *Lichenes, Jungermanniae, Marchantiae,* etc. are equally rare on the *alvar*-land and on the trees, with the sole exception of *Lichenes crustaceos leprosos* [*Lecanora calcarea* (L.) Sommf.] on the stone. The heather which grows copiously in the rest of Sweden is very rare. How fortunate the farmers would be if they could find out and imitate the manner in which Nature ousts moss and heather!

It is the custom of the inhabitants of this district to pluck the geese while they are still alive. Nature teaches them to do so, since when the ducks are sitting on their eggs, they pluck off their own down to cover the eggs and keep them warm while the mother is out looking for food. It is also known that this is the time when the geese moult and even the pinions are dropped, but when the geese are killed, in autumn, the pinions are soft and unripe. We were told that the feathers are only loosely attached and the geese do not dislike the procedure. Thus this method should be used all over Sweden, since feathers are expensive and indispensable, and the goose feathers in Sweden ought not to be wasted but collected as carefully as the sheep's wool. It should be noted, however, that the people do not pluck all the feathers from the geese; only the down and the small feathers should be plucked where the plumage is densest. A goose plucked in this way will not be completely bald, only slightly "disrobed".

Pullus Hiaticulae [young bird of the ringed plover, *Charadrius hiaticula* L.] was a little young bird, as big as a hen's egg, grey and rough with a black beak and black *lituris* [line-like markings] on the head and back, white on the neck, the breast, the belly and under the beak, grey under the throat, and with three toes to each foot.

An *Acarus* [*Sericothrombidium holosericeum* (L.), which occurred here, was entirely red (*coccineus*) with a rather small head, two eyes and eight feet, all these having many joints, a large abdomen, compressed at the head end and raised at the side with a shaggy velvet-like rear end. It could hardly be differentiated from the water mite *Acarus aquaticus ruber abdomine depresso* L. [*Hydrachna globosa* (De Geer), but when it was put in water, it floated and did not know how to dive.

The *Acarus,* which is also called *Scorpio Araneus* [*Chelifer cancroides* (L.)], walked backwards to avoid falling on its nose in spite of its having large "hands" to break the fall; the fore claws are like a scorpion's.

The crickets chirped in the house in their usual way. People told us that they were driven out by smoke from aspen, their strongest poison, where aspen is often burnt.

Here one saw the "tok" [*Potentilla fruticosa* L.], two ells high, gathered in the woods, used for besoms to sweep the floors. The leaves of this tall "tok" were smaller and narrower than those of the low and common one. The twigs were not branched so much as when they are often grazed.

Geum rivale [*G. rivale* L.] was found here in various places, *flore prolifero.* The stalks were shorter, *foliola interiora calycina simplicia, ovata, serrato-sinuata, plicata, petiolata, corolla duplo longior.* The corolla had 10–15 petals, few stamens, out of the centre in place of a pistil grew a style, where in a single [normal] nodding flower would be proper stamens and pistil.

Plants around here were *Epilobium* 1 [*E. angustifolium* L.], *Triglochin* 1 [*T. palustre* L.], *Polygonum* [*P. aviculare* L.], *Herba Britannica* [probably *Rumex obtusifolius* L.], *Viola palustris, Orchis militaris minima* [*O. ustulata* L.] and *Chamomilla foetida* [*Anthemis* 703 *Cotula* of Index = *A. cotula* L.], which grew over the whole shore.

After the church service, a man collected us together to enquire about us and our plans. He took us for spies and said that before the last war three spies had been there, who had been executed at Hulterstad; that we observed everything; investigated every situation; that the priest had given us information about the church's means; that I had all the time counselled my company with the words, that they should have eyes for everything; therefore we were advised to take with us a state attendant, which we did.

June 8 [p. 85]

In the morning, when we left the vicarage of Ås, the people of the village came running to look at us as a spectacle. The sky was getting cloudy, but the inhabitants foresaw a dry day, although the swallows were almost flying below the horses' bellies. A few hours later the rain started and continued for the rest of the day.

We travelled southwards towards the southernmost point of Öland through the groves and parks of Ottenby.

Ottenby wood is about ¼ mile of birch with some aspen and oak. In the wood are broad roads or alleys, and here and there glades and clearings. Across this stretched a small dilapidated dry-stone wall with stone shelters 2 ells high and as wide inside, about one gunshot apart, so that a man could move comfortably within one and easily shoot the deer when hunted, as they slow down before jumping over the wall. In the centre of the park there was a barn where winter provender was stored; the hay was spread on the ground within an enclosure with a low fence, and the deer would gather here whenever the gamekeeper struck the wall with his axe. Close to this house was a field where the inhabitants gather at midsummer to dance.

Midges (*Tipula thorace virescente, alis membranacei coloris, puncto nigro, vide* June 1) [*Chiromomus plumosus* (L.)] were in this wood all along to the sea, in indescribable abundance. They flew into our mouths and faces, and we were told that because of them the deer leave the woods for the fields at this time of year.

A dry-stone wall crossed the promontory at the southern border of Ottenby Park separating it from the southernmost flat part of the island, a field ¼ of a mile long. Not far from the cape was a masonry-lined well called Rosenkind's well without an outlet and with good water. Nearby, we saw the ruins of the Rosenkind chapel, which was maintained formerly by the many fishing villages here on the island in the old days when the herring passed these shores annually, but now when these have turned away, both chapel and fishing villages have declined.

Herring nets were spread out on the ground to dry. They were similar to other deep nets with meshes so small as to allow only one finger through them.

We also saw the seine used for "Tobis"-fishing here. It consisted of a sackcloth sheet, 2–3 ells [120–180 cm] long and equally broad. The net was made of horse-hair threads, thick as a goose quill, and was of the same size as the sheet with meshes wide enough to allow four fingers through. Both sides of the net were attached to the sheet. The head- and foot-ropes were longer than the net and were attached to a curved piece of wood which was used for hauling the seine. Stones were attached to the foot rope and wood chips to the head rope. The seine was not wedged, since the fish are immediately attracted by the white net; they will go for white colour wherever they see it.

"Tobis" is the local name for *Ammodytes Tobianus, Anguilla de arena* or *Sandilz ichtyologorum* [*Ammodytes tobianus* L., sand-eel] a little fish, a quarter-ell [15 cm] long, round, like a siklöja but lacking visible scales; the dorsal fin has 60 soft rays and stretches from a finger's breadth from the head along the back towards the tail; the pectoral fins have 15 rays each, the first being short, the last one the shortest; there are no ventral fins; the anal fin (*pinna ani*) from the anus to the tail has 32 rays; the cleft caudal fin has 14 rays; the head is rounded and pointed; the lower jaw is much longer and more pointed than the upper jaw; the eye is pale yellow with a black pupil; the back is grey with darkish spots across; sides and abdomen are silver coloured; the branchiostegal membrane (*membrana branchiostega*) has seven rays. The fish is used as a bait on hooks for cod fishing.

The cod caught here is of one species only. As the cod genus is a rather large one, I will describe the one caught here; the back is grey with yellowish spots; the abdomen is white with innumerable small black dots; the first dorsal fin has 14, the second has 19 rays; the first anal fin has 20, the second 16 rays; the pectoral fins have 17 rays, the ventral fins 6; the branchiostegal membrane has 7 [branchiostegal] rays. There are no more teeth than those in the jaws; a little rounded, pointed, soft barbel one finger long hangs from the tip of the lower jaw. This fish and also small herring are the principal ones to be caught here.

The shore and the southern cape of Öland showed us a great abundance of *Scandrix seminibus hispidis* [*Anthriscus caucalis* Bieb.], and *Arenaria maritima Ruppii* [possibly *Spergularia marina* (L.) Griseb.] with a purple-coloured flower grew so luxuriously that it might well suffocate itself. Among these grew *Atriplex angustissimo & longissimo folio Tournefortii* [*A. littoralis* L.] with blunt and fleshy leaves, hardly serrated at all.

We also saw *Atriplex sylvestris folio hastato s. deltoide Morrisonii* [*A. hastata* L. = *A. calotheca* (Rafn) Fries] together with *Cotula* [*Anthemis cotula* L.].

Oyster-catchers, seagulls and lapwings were flying and screeching above us. We shot an oyster-catcher (*Haematopus*) [*H. ostralegus* L.], injuring its wings so that it could not fly. The mate flew away, then returned almost immediately, reluctant to forsake the victim, but finally she had to abandon him in this predicament as he could not be helped. The bill was longer than the head, compressed from the sides, more so towards the tip, which was more yellowish; with the upper jaw slightly longer, bright red in colour; there is no other Swedish bird with so red a bill; the nostrils are oblong and one can see through them; the eyelids are featherless with the eye surrounded by a bright red circle; the eyes are red with a black pupil; the head, the neck down towards the breast and the upper half of the back from the neck are black; but the following parts were white, namely breast, abdomen, the lower half of the back and the upper and lower wing and tail coverts. The tail feathers were 12, all equally long, white from the base to the middle and black from there to the tip. The outer pinions were black but had an oblong white stripe in the middle and a white inner edge; the inner pinions were black with white margins and bases. The feet were flesh-coloured with three stout toes, the outer joined to the middle one; the claws were whitish, the thighs half naked and the tongue not cleft.

Two birds, white and unknown to us, which were slightly bigger than seagulls, were seen on land but they flew up as we approached them. One of them was soon shot, and we told that the people here call them "skärfläcka" and that they are only found at the southern cape of Öland. The mate flew out to sea, where it swam like a duck.

The "skärfläcka" (*Recurvirostra albo nigroque varia*) [*Recurvirostra avosetta* L., avocet] is altogether white except the top of the head, the upper part of the neck and the upper part of the wings, which are black. The feathers above the tail are also black. The outer pinions are the longest, of which the first eight are black at the top and white towards the base; as the white part at the base increased in size the feathers became smaller, so that the last one had only a black spot at the tip; but the other pinions were all white; the wing-coverts were white too, except the first five which were black at the tips. The tail feathers were quite short and white. The thighs were half bare, the legs bluish and two to three times longer than the tail. The feet had a little digit pointing backwards and black claws. There was a little white spot above and below the eye. The bill is the most remarkable thing about this bird; it is black, bent downwards, three times longer than the head, sharp as an awl and curved upwards, the tips of the jaws are thin and look like vellum, the nostrils are oblong and can be seen through. This bird is all the more remarkable since it is the only species to which the Creator has given a recurved bill, which it uses like a plough in the mud to find its food.

Mites are found on most birds and four-footed beasts, even also on fish. Redi has been interested enough to illustrate more than 30 species of bird mites; therefore we searched these two rare birds for mites and found their particular creatures.

The oyster-catcher mite [*Saemundssonia haematopi* (L.)] is as big as a flea, hard and tough. It ran fast and was yellow-brown (*glaucha*) in colour; the head was rounded, flat, somewhat raised; the abdomen was oval tapering towards the breast with 20 pale transverse grooves; on the sides of the back part of the abdomen were hairs; the neck was rather narrow, the feet short and the horns quite short.

The avocet mite [*Pediculus recurvirostrae* (L.)] was dark and oblong; the head was almost triangular and crossed by a furrow; the abdomen was oblong, nearly like a *linea,* somewhat broader in the middle, with eight grooves; the feet were short and curved; the horns small, short and with little buttons at the ends.

We saw several lapwing [*Vanellus vanellus* (L.)] nests in the fields, mostly with four eggs, which were greyish and unevenly spotted with dark and black dots. In one of the eggs which we opened and which had lain under the mother for some time, we saw the developing strands of life with many branched blood-red veins, a pale heart and two black eyes.

The extreme point of the southern cape of Öland could be seen with banks emerging from the water here and there, warning the seafarer against coming too close.

From here our journey proceeded northwards along the eastern coast of Öland.

After a short while we passed through the sheep pastures between Ottenby Park and the eastern seashore. This district was about 3/8 of a mile long, covering half of the grounds towards the Ottenby Estate, where sheep were bred formerly.

The plants in this meadow were above all the following:

Alopecurus culmo erecto [*A.* 52 *erectus* of Index = *A. pratensis* L.] grew where water had lain all winter and caused bare patches. *Rumex foliis lanceolato hastatis* [*R. acetosella* L.] coloured the higher patches of ground red with its flowers; *Orchis morio* grew abundantly in the meadow. "Tok" [*Potentilla fruticosa* L.] grew among the shrubs, 1½ ells tall, as a sign that the deer do not eat it as willingly as do oxen and horses. *Ulmaria* [*Filipendula ulmaria* (L.) Maxim.] grew here in the marshy spots and is nicely called "kallgräs" [cold grass] since it grows in "cold soil", i.e. a marshy soil which gives off a cold vapour during the night damaging the crops; this makes the "cold soil" spots discernible from the rest of the ground. *Trientalis* [*T. europaea* L.], *Iris* [*I. pseudacorus* L.], *Myrica* [*M. gale* L.] and [*Cepa, Allium schoenoprasum* var. *alvarense*] grew plentifully here; the last gives an onion taste to butter.

Euphorbia [*E.* 438 *oelandica* of Index = *E. palustris* L.], which Rudbeck called *Tithymalus maximus Oelandicus,* grew abundantly here in the middle of the field close to a little brook, Fällbäcken, although we had never seen it wild in Sweden before; it grew as shrubs a few ells tall; the stem perishes each year; the leaves are alternate, lanceolate, obtuse and entire. The *Involucra partialia* were 4-leaved with four flowerstalks, each branching into two stalks with two leaves; the first flowers were male with five petals, but those at the branching of the flower-stalks were hermaphrodite with four petals; the flowers were yellow and the petals entire. We saw this plant later on during the day here and there in the meadows on the *alvar*-land.

We now returned along the dry stone wall across the island, leaving the Ottenby district and following the eastern seashore northwards. We then turned westwards towards the *alvar*-land, to a farm called Kiärre, in order to see Kärrekusa (Kiärre-kusa).

Kärrekusa is a spring not far from the farm, in the bare rock, which Nature has shaped into an oblong crevice, 1 fathom long and 1 ell broad in the middle, tapering at both ends, 1½ ells deep with slightly concave sides; under the northern end of the crevice there is a hole more than a fathom deep, sloping downwards into the rock, and this well never dries up. We marvel at the shell *Cunnus marinus* and the petrifaction *Hysterolitus,* in which Nature has portrayed the female genitals; but indeed, Nature here has no less art in portraying them in all their details. One wonders if there is any *antrum* or spring in the whole world more remarkable than this one.

The *alvar*-land here was covered with tussocks and adorned with *Cistus oelandicus* [*Helianthemum oelandicum* (L.) DC.], *Anthyllis rubra* [*A. vulneraria* var. *rubra* L.] and "tok" [*Potentilla fruticosa*].

We went to see Eketorpsborg with its ruins and decayed walls which stands ¼ of a mile from the eastern shore and has once been one of the most excellent fortresses in this country; it was a musket-shot in diameter, with a spring in the middle which never dried up. Without doubt these forts have served as shelters for the inhabitants before gunpowder and shot were invented.

Stone walls, or ruined fences, enclosing vast areas of *alvar*-land, made us consider whether this barren land had once been more fertile, tilled and enclosed: since even if it had only been pasture land once, it must have been more fertile than it is now.

At present, the *alvar*-land brings forth little but *Festuca,* which I call "färegräss" [sheep grass] ; it is short here, only a finger's breadth, and grows sparsely, and is good for nothing but for the sheep, which are far fatter and healthier here than on the best grazing lands. No district in the whole of Sweden seems to be so suitable for sheep as this *alvar*-land, if only other animals would spare their winter provender.

There are a few farms in those parts of the *alvar*-land where the soil is deeper and the land higher; the lower parts of the *alvar*-land are waterlogged during winter, spring and autumn.

The *alvar*-land does not belong to any parish or village with the exception of the little farms on its borders; thus the land is not enclosed anywhere from the Ottenby wall to Borgholm castle; the horses that graze here remain in their home district.

Gräsgård inn was situated ¾ of a mile from Ottenby, but our detour made the distance much longer; the people here were robust and healthy-looking and dressed according to good old custom, in large wide trousers and with a little flap where the sleeve is attached to the bodice of the blouse.

Trebyborg lies not far from the road on the right hand; we went to see it; this fort was built on the *alvar*-land and consisted of three smaller forts in line; the first one was the smallest, the middle one the biggest, the northern one was of intermediate size; the walls were 8 ells broad; the fort was fairly strong against attack; inside the walls were ruins of chambers, where the people undoubtedly had their sleeping quarters.

At noon we visited Segerstad church. In the old parish records we read that in 1624 there had been a fire in the heather northeast of Ottenby wood, which burned so violently that only with great effort had the yeomen been able to prevent it reaching the wood; at present there was no heather here, except for a little within the wood, and it was so dwarf that one could hardly see it. Thus 100 years can change the soil without any evidence as to the cause. If conjectures are allowed in a travel account, I would guess that the heather has disappeared after the land was enclosed within the big wall, and herbs and flowers were left to grow undisturbed by the cattle; in other places I have seen how a barren heath has been turned into a meadow after having been enclosed for eight years, and I have seen vast meadows, which have been meadows for ages, covered with heather when they were laid out for grazing.

The *lantborg* and our way were less steep here, and were hardly discernible from the *alvar*-land and the seashore; the way is ¼ of a mile from the seashore.

The churches are most often built on the *lantborg* where the soil is a few ells deep, perhaps to make it possible to bury the corpses in the churchyards. The ancients probably had the same reason for choosing this site for their barrows; the *lantborg* was full of them all the way to Hulterstad. Some of these burial monuments are constructed of stones laid out in a circle, others in a square, of which the latter type were the largest and often had stone circles within them.

Milby meadow, which we passed, is ¼ of a mile broad but we did not see anything remarkable here except the *Tithymalus oelandicus* [*Euphorbia palustris* L.] and at the northern end large hawthorn trees, some of which had a crown which was 3–5 ells wide and had a trunk thicker than a man's thigh. There were no other trees.

"Tok" [*Potentilla fruticosa* L.] grew in all the tussocks which cover the *alvar*-land; further away north of Hulterstad it disappeared.

The road was like a chain of ancient barrows.

Small flour mills, which are driven with water in spring and autumn only, were seen here along the sides of the *alvar*-land; we had not seen these on Öland before. The pond, built of stone and peat, was dry now; there was seaweed (*Fucus*) stuffed between the stones, a practice which one had not seen in Sweden before, although it is well worth noticing; the seaweed does not rot easily, it swells and is so to say rejuvenated every time it becomes wet and then it is better than anything else to keep the pond watertight.

We spent the night in Hulterstad, ¼ of a mile from Segerstad.

June 9 [p. 95]

As soon as we got up this morning we walked down to the seashore, ⅛ of a mile from Hulterstad. The grass was wet with dew and the skylarks chorused in the air.

The rocks and stones lay bare at the shore, since the soil had been washed away; their eastern edges were ground down and worn by the waves of the sea.

Cartloads of seaweed (*Fucus*) were taken from the shore to the fields where it was put in heaps; one man would spread it on the field and another man ploughed it down into the soil so that only the tips were seen above the surface to protect the crops when they come forth from the grain. All those who live along the seashore have a good supply of this kind of manure, since the sea throws it up in large amounts and all the shores are covered with it. The farmers told us that in sandy fields this seaweed is soon turned into mould, although it is not a very rich manure.

All the plants that grow on the seashore turn into *plantae succulentae,* that is, plants with fleshy and juicy leaves, although the majority of them, when growing in other places, have ordinary thin and dry leaves. Such were *Atriplex angustissimo & longissimo folio* [*A. littoralis* L.], *Plantago latifolia* [*P. major* L.], *Glaux* [*G. maritima* L.], *Trifolium, Galium luteum* [*G. verum* L.], *Polygonum* [?*P. aviculare* L.]. All other succulent plants thrive in dry places and do not tolerate a wet soil. We saw no plants on the bottom of the sea, except for a *Conferva articulata capillaris ramosissima fastigiata flavescens,* which grew there like a shrub.

Cancer, Pulex fluviatilis [*C. pulex* L.] abounded here on the shore, and we examined it more closely. The tail was cleft; under the tail were two feet with two fingers each; centrally under the tail were three pairs of reflexed feet; under the body were three pairs of feet, pointing forwards with cloven claws.

An *Acarus* [*Molgus littoralis* (L.)], which was small and red and like the one found on currants in summer but almost twice as big, jumped about on the stones at the shore, quite commonly; the abdomen was reddish brown, the feet all blood-red.

Viola martia inodora [*V. stagnina* Kit.] with white flowers grew in abundance not far from the shore.

Scarabeus niger, elytris viridibus obtuse sulcatis pedibus antennisque nigris [*Carabus nitens* L.] was running on the ground; when we caught it, it spewed out its poison to revenge itself on us; this had a strong and nasty smell; its antennae were black with ten joints, almost as long as half of the body; the feet and underside were black. Head and breast were smooth and green with a yellow hue; both wing-covers were furrowed with eight furrows or seven ridges, yellowish green and shiny with orange margins; when we looked at the insect through a microscope [i.e. lens], we saw that its breast, head and the furrows on the elytra were densely covered with dots.

The blackthorn bushes here on the east side were rather small and not as luxuriant as those on the other side.

The villages here stank disgustingly of smoke, because the inhabitants had burnt dry cow's dung, which they collect from the ground on dry summer days, and for lack of other fuel they are forced to use it for baking and cooking, as on the plains of Skåne.

The outhouses were most often built of stone, for economical reasons, and few farmers could afford to build any house of wood other than his warm dwelling house. Had the people here clay of better quality and not so mealy, or mixed with sand, or else lime, no place would be better for building all houses of stone, since there is building stone everywhere as well as limestone, and the only thing needed is wood to burn the lime.

The straw roofs are covered with peat, with seaweed between straw and peat; in other places seaweed was laid on the ridge of the straw roofs without any peat.

The farmers' wives wore white head-cloths tied at the nape of the neck and the maids had plaits and a little white hood, one quarter-ell [15 cm] wide, which only covered the braided hair.

The plants used for dyeing by the farm-wives in this district were Devil's-bit Scabious (*Scabiosa succisa*) [*Succisa pratensis* Moench], which was used for green and not for yellow; Marjoram (*Origanum*) [*O. vulgare* L.] was used for brown; Bur marigold (*Bidens* [*B. tripartita* L.], not *Eupatorium* [*E. cannabinum* L.], as others have said) was used to dye woollen cloth orange.

Melampyrum spicis conicis laxis laceris [*M.* 511 *lacerum* of Index = *M. arvense* L.] grew in the cornfields with purple spikes and a straight stalk, a plant which has only been seen in Skåne before, growing among the crops.

Pulsatilla, foliis decompositis pinnatis, flore pendulo [*P.* 447 *retroflexa* of Index = *P. pratensis* (L.)] grew on all bare and dry patches and was avoided by horses. Wormwood grew abundantly in all the villages.

Ranunculus foliis ternatis integerrimis [*R.* 471 *illyricus* of Index = *R. illyricus* L.], of which we saw a few plants on the fifth of June at Bårbyborg and which not only has never before been seen in Sweden but is also rather rare in the rest of Europe, was found in the cornfields and on the headlands between Hulterstad church and the next village, to the north, along the road, in incredible abundance. It had several leaves and branches on the stem. The flowers are yellow and glossy, larger than in *Ranunculus acris.*

Squamula nectarii [nectary scale] at the base of each petal was crenate and *pistillorum capitulum* [head of carpels] was oblong as in *Adonis; radix tuberosa sessilis conglobata* had long fibres projecting from the sides, and at the end of these were little *radices tuberosae.* The whole plant was tomentose like *Cyanus* [*Centaurea cyanus*]. The stem was straight and half an ell [30 cm] tall. The root leaves were ovate-lanceolate and obtuse, with stalks. There were three stalked leaflets on each petiole. Of these, the lowest one was slightly longer and the upper ones narrower. The flowers at the top branches had five sepals each. Some flowers had white spots, but on these the upper membrane was separated from the rest.

In order to see Tribärgaborg we left the main road at the first hamlet and travelled to the left towards the *alvar*-land; it was a musket-shot [about 225 m] in diameter; its walls were quite strong, lined and surrounded by another wall; a few fathoms from this castle we saw the ruins of a wall, doubtless a fortification. We broke away a few stones to see if it had been walled with mortar, and we found small *vestigia* thereof although the many centuries had almost obliterated them; we found small *cochleas* [snail shells] with 7 twists, as big as a wheat grain, brownish, spiralled counterclockwise; *Buccinum exiguum flavum mucrone obtuso s. cylindraceum. Lister. angl. 121. t. 2. f. 6.* [*Pupilla muscorum* (L.)].

The farmers are not allowed to carry guns here, to prevent them from harming the deer; in their own opinion they should have been ordered to have them, especially in times of war, since they believed that they would be able to ward off such small vessels as are able to land on their shores. A few Finnish boats had anchored in the roadstead a few days ago, and the inhabitants gathered armed but had to return disconcerted, when it was their own people that they met.

The *lantborg* could hardly be seen here; it was not very much higher than the rest of the land and not steep and lay 1/8 to ¼ mile from the sea.

We passed Stenåsa church; the fields were yellow with charlock.

We arrived at Sandby church at noon. In the graveyard were two tall runic stones with snake-like ornamentation nicely twisted like double knots; on the stone closest to the church we read the following letters:

ᛚᚦᚠᛆᛋᛏᚱ×ᛆᚢᚴ×ᚼᛁᛚᚴᚢᚾ×ᛆᚢᚴ×ᛁᛂᚾᛁᛁ×ᚦᛆᚢᚾ×

The second line:

ᛘᚢᚦᚴᛁᚾᛁ×ᚴᛁᛏᚾ×ᚱᛂᛁᛋᛏ×ᛋᛏᛂᛁᚾ×ᛁᚠᛏᚱ×
ᛁᚢᛂᛁᚾ×ᚠᛆᚦᚢᚱ×

Linnaeus: *lþfastr* × *auk* × *helkun* × *auk* × *ienii* × *þaun* ×
muþkini × *kitn*×*reist* × *stein* × *iftr* ×
iuein × *faþur* ×

Öl 26: × *kuþfastr* × *auk* × *helgun* × *auk* × *nenir* × *þaun* × *myþkini* × *litu* ×
reisa × *stein* × *eftir* × *suein* × *faþur*

"Gudfast and Helgunn and Nenne, mother and sons, had this stone raised in memory of the father, Sven."

The other stone carried the following inscription:

ᚦᛆᛦ×ᛒᚱᚢᚦᚱ×ᚱᚽᛁᛋᛏᚢ×ᛚᚢᛒᛚ×ᚦᛁᛆ×ᛆᚠᛏᛁᚱ×
ᛋᛚ··ᛏᚢᛚ:×ᛋᛁᚾᛆ×ᛆᚠᚱᛁᚦᛁ×ᛆᚢᚴ·ᚽᚠᛏᛦ×ᛋᚢᚽᛁᚾ×
ᚠᛆᚦᚢᚱ×ᛋᛁᚾ×ᚠᚮᛁᚾᛆ×

Linnaeus: *þaʀ* x *bruþr* x *reistu* x *lubl* x *þia* x *eftiʀ* x
sl . tul: x *sina* x *afriþi* x *auk* x *eftʀ* x *suein* x
faþur x *sin* x *foina* x

Öl 27: x *beiʀ* x *bryþr* x *reistu* x *kubl* x *þia* x *eftiʀ* x *sustur* x
sina x *afriþi* x *auk* x *eftʀ* x *suein* x *faþur* x *sin* x *koþan* x

"Those brothers raised this memorial in memory of their sister Åfrid and in memory of Sven, their good father."

It could easily be seen that both stones were carved by a master, since in both there was a strange letter, probably an F, consisting of a perpendicular line with a pit at the middle which was cylindrical and large enough to hold the little finger.

A child was brought to me by a poor woman. She complained bitterly that her neighbour had reproached her for giving the child a man's name although it was a girl; she asked me to tell what sex it was; when the child was examined we found that the penis was not visible, but there was no oblong rima but a rounded hole, and in this was a *caruncula* as big as the tip of the little finger. When the *caruncula* was bent down, the *glans penis* appeared. The urine was passed from above the *glans,* and there was a scrotum with two testicles below, as in a male child.

The plants used for dyeing in this district were: fresh leaves of *Scabiosa succisa* [*Succisa pratensis* Moench], which were boiled together with the yarn laid layer upon layer (*stratum supra stratum*) for as long a time as it takes to cook fish; it is left overnight in the cauldron, and in the morning the yarn is removed; it is then still colourless; the cauldron is reheated, and the yarn is hung on rods over the cauldron and covered, so that the vapour saturates the yarn and gives it the green colour; when the yarn is removed from the cauldron, it is wrung out, the leaves are thrown away, ashes are added to the decoction and the yarn is often dipped into it. Bur-marigold [*Bidens tripartita* L.] is used thus: the yarn is washed, dried, mordanted with alum, and dried once more; the fresh green marigold is chopped, mixed with the yarn in layers, boiled for a couple of hours, and stirred well to give a uniform yellow colour. "Stone moss" [*Parmelia saxatilis* (L.) Ach.] is used for brown, mixed layer by layer with the yarn, which is boiled in water with a little lye* to fasten the colour, but the yarn is not mordanted. For light yellow, Juniper moss (*Lichen fulvus, sinibus daedaleis*) [*Cetraria juniperina* Ach.] is used; it yields a beautiful yellow colour. *Serratula* [*S. tinctoria* L.] also gives a yellow colour when the yarn is mordanted with alum; the weed is chopped, boiled layer by layer and,

* Lye is a strong alkaline solution made by soaking wood ash in water.

when it has boiled, a little dye is added and the boiling continued. *Origanum* [*O. vulgare* L.] gives an intense brownish red colour; the plant is dried, chopped and boiled with the yarn, which is often lifted out and pounded. Apple-tree bark yields a lemon colour, brighter than *Serratula.* Soot boiled with water gives a brown colour, darker than the stone moss.

Sempervivum [*S. tectorum* L.] grew on the roofs, *Athamanta libanotis dicta* [*A.* 229 *campestris* of Index = *A. libanotis* L. = *Seseli libanotis* (L.) Koch], *Onopordum* [*O. acanthium* L.] and wormwood grew in the churchyard; there was elder in most hamlets and *Cichorium* [*C. intybus* L.] and *Behen album* [*Silene cucubalus* Wibel] along the fields.

The kind of seaweed called *slak* [*Fucus vesiculosus* L.] is bought from Torslunda and is used for thatched roofs. It is much more suitable than ordinary seaweed, *ex. relat. pastoris loci* [according to the local clergyman].

Herring and cod are fished on the eastern shore, but not in the west; we saw dozens of farmers coming to buy fish.

Gårby church was situated ¼ of a mile from Sandön. In the graveyard we read the following letters on a runic stone:

ᚼᛆᚱᚦᚱᚢᚦᚱ×ᚱᛁᛋᛁ×ᛋᛏᛆᛁᚾ×ᚦᛁᛁᛋᛁ×ᛆᛁᚠᛏᛁᛦ×
ᛋᚢᚾ×ᛋᛁᚾ×ᛋᛁ·ᛁᚦ·ᛏᚱᛆᚴ× ᚠᚢᚦᛆ××ᚼᛆᛚᚠᚿᛏᚱᛁᚾ×
ᛒᚱᚢᚦᛁᛦ×ᛆᚾᛋ×ᛋᛁᛏᚱᚴᛆᚱ.

In the centre of the stone we saw the following letters, which were smaller and more damaged.

ᛏᚱᛁᛆᚴᚦᚢᛁᛁᚱᛆᚦᛆ×ᚴᛆᚾ·

Linnaeus: *harþruþr* × *risi* × *stain* × *þiisi* × *aiftiʀ*
sun × *sin* × *si.i.þ trak* × *fuþa* × *halfntrin* ×
bruþiʀ × *ans* × *sitrkar*

triakþuiiraþa × *kan*

Öl 28: *harþruþr* + *raisti* × *stain* + *þinsa* + *aiftiʀ* × *sun* + *sin* +
smiþ × *trak* + *kuþan* + *halfburin* + *bruþiʀ ans* × *sitr* × *karþum*
brantr × *rit-* × *iak þu raþa* + *kan*

"Härtrud raised this stone in memory of Smed, his son, a very good man. Halvboren, his brother, sits in Gardarike (i.e. Russia). Brand hewed properly so that you can read."

Norra Möckelby was situated ½ a mile from Gårby, but since we saw nothing remarkable, we continued to Runsten Church, ½ a mile's journey.

On our way to Runsten, we saw the same type of mound as in Lapperstad, and in a village near Runsten a rather large number of square upright stones.

This night we rested after a hot day in Runsten, with the vicar Magister Anders Brauner, who accompanied us with much consideration the following day to show us the interesting things in the district.

June 10 [p. 103]

Runsten church has a runic stone with a short inscription at the southern wall, as follows:

ᚾᛁᛅᛚᛒᛁ·ᛁᛅᛚᚢᚾᛅ·
ᛅᚢᚦᛒᛁᛅᚱᚾᛚ · ᛁᛏᚱᛁᛁ'ᛁ'ᛏᛁᛁᚾ ·

Linnaeus: *nialbi . ialuna .*
auþbiarnl . itreisastein .

Öl 31: *auþbiarn . lit : reisa : stein : . . . hialbi : sialu . has*

"Ödbjörn had (this) stone raised . . . may help his soul."

It is the custom here for brides, on the wedding day, to run to the church together with the bridesmaids. The bridegroom rides there with his men, and the bride has to make sure that she arrives soon after him. Before the wedding day, on the third announcement of the marriage, she wears a cloak when leaving the church. She is accompanied by a maiden old enough to be married, and they visit the parishioners, one after another. She enters the house and stands motionless inside the door and bows without uttering a word; the language is however understood, one gives her plates, another saucers, a third one spoons, and so on, and thus she receives everything she needs for her household work. The housewives have always prepared their gifts beforehand, and the maid puts them in her bag and helps the bride to carry them. The supplicant has often traversed the whole village before the churchgoers have left the church, since the parishes are small and consist of only a few hamlets close to each other, usually all within ¼ to ½ a mile.

The compass is called "Norrigen" here, and a hand-mill is "Pinan".

We had hardly left Runsten Church when we saw at Bjärby a big upright runic stone, to the right of the road, with three circles of text-bands:

α ᛁᚢ·ᚱᚭᛚᛁᛏ·ᚱᛅᛁᛋᛅ·ᛋᛏᛅᛁᚾ·ᚦᛁᛋᛅ·ᛁᚠᛏᚠᛏᚱ'ᚠᛁ ᚠᛅᚢᚱ
β ᚱᛅᛁᛋᛁ·ᛅᚠᛏᛁᚱ·ᚢᚿᛏᛅᛋᛁᛋ ᚼᛅᚢ ᛁᛅᛦ ᚠᚱᛅᚠᛁᛅᛁᚴᛁᚱᛁᚴᛁᚢ
γ ᚢᚠᛚᛁᛁ··

Linnaeus: *iu . rolit . raisa . stain . þisa . iftftr-fiaur*
raisi . aftir . untasisha-iaʀ frafiaikirikiu
uflii

Öl 36: × *harfriþr* × *auk* × *uiþbiurn* × *litu* × *raisa* × *stain* × *þina* × *at* ×
fastulf × *faþur sin* × *siklauk* × *lit* × *raisa aftiʀ* : *bunta* × *sin* ×
han × *iaʀ* × *kraftin* × *i* × *kirikiu* ×

"Härfrid and Vidbjörn has this stone raised in memory of Fastulf, their father. Siglög had (this stone) raised in memory of her husband. He is buried in the church."

Soon afterwards but before we reached Lerkaka we saw another runic stone, in two pieces, with two text-bands:

ᛁᛅᚠᚱᛅᚢᚴ·ᚴᛅᛘᛅᛚᛅᚢᚴ×ᛋᛅᚴᛋᛁ×ᚱᛅᛁᛁᛏᚢᛋᛏᛅᛁᛁᚬ×
ᚦᛁᚼᛅ×ᛁᚠᛏᛁᛦ×ᚢᛁᚠᛅᚦᚢᚱ×ᛋᛁᚾ×ᚴᛅᛁᛁᚱ····×ᚴᛏ×
ᛒᚬᚾᛏᚾᛋ·ᛋᛁᚾ×ᛁᛅᚢᛋᚢ·
ᚴᚢᚱᚦᛋᛁᚠᛁᛅᛦᛁᚢᚾᚬᛚᛅᚠᚾᚼᛅ ᚠᛅᛏᛚᛁᛏᛘᛁᛘᚢᛅᛁᚢᚾᚾᚦᛅ
ᚱᚼᛅᛚᚠᛅᚾᛒᚢ·

Linnaeus: *iafrauk . kamalauk* × *saksi* × *raiitustaiio* ×
þiha × *iftiʀ* × *uifaþur* × *sin* × *kaiir . . .* × *kt* ×
bontns . sin × *iausu .*
kurþsifiaʀiunolafnhe fatlitmimuaiunnþa
rhalfanbu .

Öl 37: *olafʀ* × *auk* × *kamal* × *auk* × *sagsi* × *raistu* × *stain* × *þina* × *aftiʀ* × *un* × *faþur* × *sin* × *kaiʀui* × *lit* × *at* × *bonta* × *sin* × *kiarsuk kubl þsi .*
fiaʀun olafʀ hefnti at miomu ati un hiar halfan by

"Olof and Gammal and Saxe raised this stone in memory of Unn, their father. Gervi had this memorial made here after her husband. Wealthy-Unn was revenged by Olof at *miomu.* He owned half the farm."

A round stone, a globular greystone, slightly flattened (*lenticulari globosus*), 3½ fathoms in circumference, lay on top of a cairn in a meadow at the road, opposite Folkslunda, ¼ of a mile from Runsten Church. It was said of this stone that it lay exactly in the middle of the island, between the southern and northern ends, and thus we were in the centre of the land, to the east, since the shape of the island is oblong.

Ranunculus seminibus aculeatis, foliis superioribus decompositis linearibus [*R.* 470 *echinatus* of Index = *R. arvensis* L.] was common in the cornfields. The root perishes each year; the stem is a quarter-ell [15 cm] long, the radical leaves are three-pointed; the lower stem leaves are three-parted, but the upper ones are nine-parted. The scales of the nectaries on the petals are obtuse, the seeds compressed, attached without a *receptaculum* to the centre of the flower, prickly on the sides and pointed at the top.

We arrived at Längelöt Church, ½ a mile from Runsten; here we left the road and travelled westwards to see Ismanstorpsborg, and soon we saw Borgholm Castle; the landscape changed greatly here; it was full of juniper, small hazel, birch and oak, and hawthorns became everywhere more common; the hazel had very good nuts but was not very tall, often less than 3 quarter-ells [45 cm].

Three-eighths of a mile from Långelöt Church we found Ismanstorpsborg, in the middle of the wood, overgrown with trees.

Ismanstorpsborg lies ½ a mile from both the eastern and the western shores, in the centre of the land, and equally distant from the north and south ends. In comparison to the other ruins seen on the *alvar*-land, this was like a capital among hamlets. The walls were still more than 2 fathoms [3.6 m, 12 feet] high and 3 fathoms [5.4 m 18 feet] wide, but many stones had fallen down. The diameter was 150-160 ells [90-93 m, 300-320 feet], partitioned with small dilapidated walls, like in a monastery. There were rooms in all directions, but the plan could not be clearly discerned, since the ruin was so overgrown with bushes and trees. The rooms to the sides of the walls were like little rectangular chambers, and between each pair of chambers there was as much space as the width of the chamber. In the centre of the ruin the chambers were arranged in a less orderly fashion. In this ruin grew *Rhamnus catharticus, Ossea* [*Cornus sanguinea* L.], *Sanicula* [*S. europaea* L.], *Asclepias* [*Vincetoxicum hirundinaria* Medicus], *Mercurialis* [*M. perennis* L.], *Trichomanes* [*Asplenium trichomanes* L.], *Hypericum* [*H. perforatum* L.], *Polygonatum* [*P. odoratum* (Miller) Druce] and *Orchis militaris minima* [*O. ustulata* L.].

We rode along the same road back to Långelöt Church.

We saw no mineral springs in this part of the country until we came to Långelöt Church, where, in the churchyard, east of the church, were two springs of which the lower one close to the wall had mineral water. The inky taste, the blue scum on the water and the yellow ochre outside the wall caused us to make the following tests:
α *Saccharo Saturni* [sugar of lead, lead acetate] gave the water a milky colour; β *Resina Gallarum* [resin of oak-apple galls] coloured it brown; γ *Solutione Lunae* [silver nitrate solution] turned it opalescent and later slightly brownish; δ *Spiritus Salis amoniaci* [sal ammoniac] made no change; ϵ with *Succo Heliotropii* [sap of *Chrozophora tinctoria*] it was coloured violet; ϛ *Minium* [red lead] removed the inky taste. This water could be used by sick people with several sorts of malady, if only the springs were cleaned, and provided one is not too squeamish about ingesting minerals from corpses and dead bodies.

From here we proceeded on our journey to the right towards a little grove and some cornfields. This place was well known for its rare flowers, but we saw nothing that we had not seen before: *Orchis hiante cucullo* [*O. militaris* L.] grew here, disdainfully common. *Orchis militaris minima* [*O. ustulata* L.], *Orchides palmatae omnes* [*Dactylorhiza* species], *Origanum* [*O. vulgare* L.], *Viscaria* [*Lychnis viscaria* L.], *Turritis I foliis subtus ferrugineis seu leprosis* [*T.* 544 *glastifolia* of Index = *T. glabra* L.], *Polygonatum* [*P. odoratum* (Miller) Druce], *Geranium pedunculis unifloris* [*G. sanguineum* L.], *Jacea laciniata* [*Centaurea scabiosa* L.], *Muscipula montana hirsuta Rupp* [*Silene nutans* L.], all were common here.

The walls here were caulked with bladderwrack instead of moss and, still better, with grasswrack.

"Gräsja" is the local name for the long narrow leaves, like long ribbons, which are thrown up on the shore by the sea. They are either all black or all white. My conjecture is that they are leaves of *Ruppia* [*R. maritima* L.] or more likely *Alga angustifolia vitriariorum I. B. : C.B.* [*Zostera marina* L.], which are black at first but turn white when bleached by the sun, *Raj. syn. 3.p.52.n.1.* He who finds a fruit among these leaves will more confidently determine their genus than we, who were unable to find any.

On our continued journey we entered a hamlet where the floors were sprinkled with *Ulmaria* [*Filipendula ulmaria* (L.) Maxim.] and double narcissi.

A child was brought to us, who was considered to be a changeling. He was a

13-year-old boy, misbegotten and without wits; he could not sit or stand up or walk; his hands and legs did not obey his mind; he could not talk but only mumble; he had protruding front teeth and a squint; he appeared to be deaf; his legs were thin, his flesh loose and he looked more like a girl than a boy; his movements were so ungraceful that they made us shiver. He was born to unhappiness, being of no delight to himself nor of no use to the public, and no praise to his Creator; but others would learn from him to thank their Creator for a well-shaped body. He was no changeling but suffered from a disease called *Hieranosos.* I asked the mother if she had been frightened during the time she was pregnant with him, and she told me that she had been terrified by two men who were fighting with knives and nearly killed each other; after this she noticed that the foetus was lying motionless on its side, as if it were dead, and not moving like her other children.

We continued our journey from Långelöt to Gärdslösa church; on the way we found a runic stone at the road in upper Bärby on our right, which was so covered with moss (*Lichene crustaceo leproso albo*) [*Lecanora calcarea* (L.) Sommerf.] that we could only read a few letters:

ᛅᚢᛅᛚᚱ·ᛅᚢᚴ·ᛅᚦᚢᛅᛏᚱ·········ᛅᚦᛁᚾᛅ·ᛁᚿ·ᚴᛁᚱᛁᚦ·

Linnaeus: *aualr . auk aþuatr . . . aþina . in . kiriþ*

Öl 41: *austain . auk . sahuatr . auk . aþuatr . litu . raisa . stain . þina . auk . bru . þsa . aftʀ . ayi . faþur . sin kuþan*

"Östen and Sigvat and Advat (?) had this stone raised and (had) this bridge made in memory of Ave (?) their good father."

A little distance further on before we came to Gärdslösa we saw two upright stones by the millpond. The northern stone had no inscription; on the southern one we read the following:

ᚢᛅᛁᛁᛁ·ᛁᛅᚱᚦᛚ×ᚽᚠᛏᛁᛦ×ᛁᚿ×ᚠᛅᚦᚢᚱ×ᚢᛁᚴᛅᚱ×ᛌᚢᚿ×
ᛅᚦᚿᛁᚴᛁ ᛁᛁᛅᛚᚠᛦ×ᚱᛅᛁᛋᛏᛁ×ᛁᛏᛅᛁ·

Linnaeus: *uaiii . iarþl* × *eftiʀ* × *in* × *faþur* × *uikar* × *sun* ×
aþniki iialfʀ × *raisti* × *itai .*

Öl 39: × *suain* × *kiarþi* × *iftiʀ* × *sin* × *faþur* × *uikar* × *sun* ×
ainiki × *sialfʀ* × *raisti* × *stain*

"Sven made (this memorial) after his father Vikar. The only son raised the stone himself."

The women here were always busy with something; most often they were knitting stockings.

The floor stones at Gärdslösa, which were of a reddish Öland stone, were all wet on this hot summer day, which was a foreboding of rain, and the rain fell the next day; it is a strange feature of this stone, which has much puzzled many naturalists. This peculiarity is probably the reason why people here do not build their dwelling houses of this stone, for, in spite of it being available and suitable in other respects, it would undoubtedly be harmful to health.

We spent the night in Gärdslösa after a warm and delightful day.

June 11 [p. 109]

When we left Gärdslösa in the morning, we were accompanied by the vicar, Mr Lundvall, and the local sheriff, since they had been informed that the inhabitants suspected us of being spies and watched carefully over our movements.

Before we came to Gärdslösa church we saw four stone boulders in a brook, the third of which had been a runic stone but it was so abraded that only the ornamental serpents could be seen.

Plantago foliis semicylindraceis integerrimis [*P.* 127 *anguina* of Index = *P. maritima* L.] covered the entire field towards Tjusta Village; the leaves were spotted, as on a serpent, and where they came up from the root there was long white wool.

The best and most beautiful wheat not only of Oland, but perhaps of all Sweden, grows here in Gärdslösa, which is the largest parish in the country.

We noticed that the soil was more sandy here, and in the rest of the northern part of Öland; hence it is more prone to suffer from drought, and the inhabitants have to add seaweed to the soil to prevent the strong wind from removing the sand and leaving the roots bare; after sowing, the farmer covers the entire fields with seaweed when he has time to do it.

We could hardly discern the *lantborg* today, since the slope is very gentle from the *alvar*-land to the seashore; the land between the *alvar* and the sea is, however, covered with stone and gravel.

At last we reached Bredsättra church, which is the annex of Gärdslösa; from here we could see the chapel at Kläppinge ¼ of a mile to the east on the point of a promontory, and we proceeded through the beautiful fields of crops.

Charlock [*Sinapis arvensis* L.] covered all fields, which made them yellow and gave a pleasant smell; the farmers told us that the charlock seed contaminates the grain, and gives it a bitter taste; thus some people remove them by riddling.

Seaweed (*Fucus maritimus, s. Quercus maritima, vesiculas habens C.B.*) [*F. vesiculosus* L.] covered all shores and, when we reached the shore, it gave off such a stench, as if ten horses had been roasted in the sun, that we could hardly endure it. There was a red colour, like blood, on the seaweed, extracted by sun, water and putrefaction; it does not colour linen, but whether, and in what way it colours wool is not known. I have heard that the French collect a *Fucus* from the bottom of the sea which they use to heighten the red colour of their wine, but it is not the same as ours; time, occasion, chemistry and experiments will probably one day imitate this operation of nature and derive some use from it.

Kläppinge chapel far out on the islet had undoubtedly been a fishing village once; nothing was left of the church except the high walls and the two gables, and instead of a floor there was *Scandix seminibus hispidis* [*Anthriscus caucalis* Bieb.].

East of the chapel in the graveyard was a well with fresh-water. To the south in the direction of the shore there was a cross, hewn out of solid stone, as high as a man on horseback. Around it were several little pits, remnants of fortifications and inhabitations.

We had occasion to see here the lump sucker fish (*Cyclopterus s. Lumpus*). [*Cyclopterus lumpus* L.], which is said to be rarely caught and then only when a storm is approaching; it is similar to a perch; the forehead is square with scales; the body is bare with seven lines of large longitudinal scales, as on a sturgeon; the scales have sharp edges; the abdomen is large; the foremost back fin is [present only in a juvenile fish, which Linnaeus evidently had] covered with warty scales and has six soft rays, the other back fin has 11 rays, the anal fin has ten rays; the two latter are opposite to each other; the tail is not cloven and has ten rays, the pectoral fins have 20 rays.

A fish called "kuder" was mentioned here and in other places, but we had no occasion to see it; according to the description it was identical with the *ålekusa* in Småland [*Zoaces viviparus* (L.), the eelpout].

Hemp and flax were seen in a few places; these are uncommon here and indeed not common elsewhere; when we asked the inhabitants why they do not grow flax and hemp, the answer was most often that they had no suitable place to ret it in.

We passed Eby church, which is an annex.

We saw Löt church in the evening; we were warm from the bright sunshine and rested a few hours there during sudden and intense rain. Nothing is so difficult for travellers here at present as to obtain fresh water, for *pretio* or *precibus.*

There was a little island here, close to the mainland, but a strong south-western wind prevented our going there, and yet it was only at a gunshot [225 m] distance from the shore.

Lillholmen is a promontory $\frac{1}{8}$ of a mile [1.3 km] long and ½ a gunshot [about 110 m] broad, which was reached after following the shore northwards for some time. This island appeared very suitable for growing woad [*Isatis tinctoria* L.]; the land is low and has no hills except a little one at one end of the island, the sea wind blows over it continuously, and the shores are flooded in the winter; the soil is suitable and fertilized by the seaweed which is annually thrown up by the sea. Thus this islet has all prerequisites for woad culture and I doubt if there is a more suitable place for it in Sweden.

Artemisia foliis compositis multifidis tomentosis, ramis floriferis nutantibus [*A.* 671 *Seriphium* of Index = *A. maritima* L.], also called *Absinthium Seriphium,* grew so abundantly on the islet as to colour it all white; it smelled so pleasing and strong that it competed with *Marum* [*Origanum*]; the leaves were *pinnata: pinnis ramosissimis, oblongis, subteretibus:* the stem is quarter-ell [15 cm] long and white with heavy and short wool; it would be well worthwhile to grow it in gardens for its beautiful scent; I have only seen it in Skåne before.

Cochlearia Danica repens C.B. [*C.* 538 *minor* of Index = *C. danica* L.] grew here on the shore quite commonly with its pentagonal leaves and smooth pods; this plant is so well known for its power against scurvy that there is no need to go into detail.

Scirpus culmo triquetro, panicula conglobata foliacea, spicularum squamis trifidis: intermedio subulato [*S.* 39 *maritimus* of Index = *S. maritimus* L.] grew in the sea close to the shore.

From here our journey was continued to Pesnäs; we passed a meadow where *Tithymalus oelandicus Rudb.* [*Euphorbia palustris* L.] grew and not far thereafter a hamlet called Husvalla.

Anemone seminibus hirsutis [*A.* 449 *major* of Index = *A. sylvestris* L.], which has not been seen in Sweden before, we found growing among the stones at the edges of the cornfields, at Husvalla, immediately before we came to the village and there in abundance. This pleasing and large flower ought to be generally planted in gardens especially since it flowers for a long time. This anemone is similar to the wood anemone, but stem, leaves and flower are three times as big; the wood anemone is glabrous, but in this one the stem, leaf-stalks and the underside of the leaves and the petals are bestrewn with white hairs; the seeds of the wood anemone are glabrous, but in this one hairy; the receptacle in the wood anemone is globose, but in this one oblong and hairy; this has always five petals to its flower, which are ovate and often notched at the end (*emarginata*), but the wood anemone has usually more petals, which are more deeply notched; the leaves on the stem are more deeply cut in this than in the other.

Matkroken is the name of a curved reef ½ a mile long close to Husvalla which has been fatal for many seafarers, since it consists of rock and stone. We asked the inhabitants why it was called Matkroken (The Food Hook); some answered that it had to have some name; others said that it was because of its shape, and also because of the advantage it gave to the inhabitants formerly, when the shipwrecks were plundered perhaps by the farmers and the sailors killed.

The journey continued; we saw many marshes alongside the road; a strong S.W. wind blew and rain fell now and then, until late in the evening we reached Södvik inn, where we spent the night, ¼ of a mile from Pesnäs.

June 12 [p. 113]

This day's journey was rather short, since in Pesnäs we heard that the yacht which sails between Öland and Gotland arrives here on Tuesdays only.

Pesnäs Church stands in a sandy district; the vicarage was better built than any other house here and it was adorned with a garden, which is rare, and there were also other fine buildings resulting from the inspiration of the vicar, Lars Hök.

The crop grown here is mostly rye; there is little barley since the sandy soil is too hard when there is a drought; in contrast to this is Mörbylånga where we were on the west side (June 5), which is the best place for crops in all Öland, wet or dry weather notwithstanding.

Stags keep mostly to the northern districts, hinds to the southern, for which reason one must everywhere have watchers in the fields, to scare off the deer and look after the protective walls.

Wild swine are most common in the northern district and do the farmers much harm, since they dig up the ground, like a plough, and trample down the crops; they walk about and beat the crops together, like sheaves, so that they can fill their mouths with ears at each bite. Sometimes they mate with the domesticated swine, which produce wilder offspring.

Wood has to be bought from Småland here too and is paid for with grain rather than money; thus each cord (or armful) will cost the yeoman about 1 *daler*, and those who transport it are often delayed by contrary winds.

The farmers eat very well on Öland, wherefore they become heavy and strong; they often use slices of ram's horn for the soles of their shoes, which makes them last longer. The womenfolk were making stockings all the time, and the wenches were skilfully making embroidery work for pillows and the like with wool of many colours; their shifts were sleeveless and had only a small band over each shoulder. Wool was dyed with *kasa* or with the spikes of reed; the local name for pepper is "fränt" and for yeast "krydda".

The soil is very dry, since under it is a fine white sand called quicksand, although it does not move, and the farmers have to beware of ploughing too deeply in case they touch the sand.

The most characteristic plants here are: *Linum catharticum, Ranunculus echinatus* [*R. arvensis* L.], *Cynosurus bracteis pennatifidis* [*C. cristatus* L.], *Alopecurus culmo erecto* [*A. pratensis* L.], *Alopecurus culmo infracto* [*A. geniculatus* L.], *Polygonatum pedunculus multifloris* [*P. multiflorum* (L.) All.], *Libanotis Riv* [*Seseli libanotis* (L.) Koch], *Sanicula* [*S. europaea* L.], *Melica* [*M. nutans* L.] varying with a *panicula ramosa* [*M. uniflora* Retz.], *Dulcamara* [*Solanum dulcamara* L.], *Orchis alba bifolia* [*Platanthera bifolia* (L.) L.C. Rich.], *Orchis hiante cucullo* [*O. militaris* L.], *Orchis militaris minima* [*O. ustulata* L.], *Orchis palmata longis calcaribus* [*Gymnadenia conopsea* (L.) R. Br.] sometimes with *flore albo, Scorzonera* [*S. humilis* L.] often with a stem ½ ell [30 cm] high, *Sedum tridactylites* [*Saxifraga tridactylites* L.].

Orchis militaris minima [*O. ustulata* L.] is here called "krutbrännare" (burning gunpowder), an incomparably appropriate name, since the flowers in the spike are burning red, but the little buds in the top are black, like a picture of a fire.

Campanula hispida, floribus sessilibus, capitulo terminatrici, foliis lanceolato-linearibus undatis [*C.* 183 *altissima* of Index = *C. cervicaria* L.]. This bellflower, which is very rare in the rest of Sweden, was common in the meadows here; the stem was fairly tall, without branches, unevenly pentagonal and bestrewn with transparent, reflexed bristles; the leaves were *lineari-lanceolata* with wavy margins.

Asperula Rubia Cynanchica dicta [*Galium triandrum* Hylander] grew abundantly in the meadows; it has red bushy roots, which should be tried for dyeing, since the plant is closely related to *Galium album* [*G. boreale* L.] and *Rubia* [*R. tinctorum* L.].

Equisetum [*E. arvense* L.] or foxtail grew in the meadows over all Öland, but especially here in great abundance, which is the sole difficulty in keeping sheep on Öland, since the English have noticed that, if sheep eat too much of it in spring, they abort their foetuses.

Quarrying has been started in the eastern part of the island too, in the parishes of Pesnäs and Föra. We noticed here that the floorstones quarried in the *alvar*-land are better in dry rooms, while the stone derived from the sea withstands water better; the stone that is quarried close to the sea becomes humid in wet weather, and sparks and bursts more in a fire. The floorstones were of four sizes: *Alnsten* or 3 quarter-ell stone, 3 quarter-ells [45 cm] long and broad; *Kronalnsten,* 4 quarter-ells [60 cm] long and broad; *Finnalnsten* 1½ quarter-ells [22 cm] long and broad, thus the smallest; and *Sexhuggare,* 6 quarter-ells [90 cm] long and 3 quarter-ells [45 cm] broad. The quarriers cut the stone in a very precise manner, according to previously laid out markings, so that the stone bursts along the predetermined lines after being split with the chisel; the transverse cuttings were made against the knees or the thighs which was called "couper". The stone chips were used for brick kilns and fireplaces; a lime was burnt from them, which is suitable for building material although it does not turn white; especially good lime is made in Trosnäs. The stone dust is harmful to the eyes, and those who are much exposed to it have red eyes and often finally go blind. When the stone has been ground, sand is added, the stone is abraded and a powder is formed; this is mixed with water and spread around the lower stone ring. The quarriers call it "sturleer" (scouring sand) and use it for masonry instead of mortar; it functions like lime and turns hard in spite of its being unburnt. Thus the lime is reduced to its *primum principium*, clay. Perhaps this mortar could be used as cement for underwater masonry, either as it is or with some addition. This ought to be tried. Small stone chips are laid at the eaves here like birchbark in the rest of Sweden, and only a little bark is needed where the stones touch each other to make it impenetrable to water. We also noticed that the workers throw the little soil there is on the *alvar*-land among the stone chips, where it disappears; however Nature obliges them, when they destroy the land for all time to come, to save the soil for the thin grass and thus improve the pasture; as the *alvar*-land is treated now, it will soon be only stone and of no use for the sheep, for which it now seems to be created. To forbid the farmers to destroy the *alvar* would be to deny many a way of obtaining their living, but there ought to be some middle way with this quarrying, since many of those who are too eager about grinding and quarrying neglect their farms and produce too much stone, which lowers the price.

We did not see any particularly rare insects today, only the following: *Aphis* [*A. cirsii* L.], dark, with six rows of elevated spots along the back; the wing buds, the rear horns, the feet and the true horns black; the thighs white towards the base. *Musca,* altogether pale with green eyes and four lines of black dots on the abdomen, with five dots in each line; white wings with four dark transverse lines. *Phalaena,* rounded with bristles and a double transverse line like a semicircle over the joined wings.

There was a garden at the vicarage here. When we asked why gardens are so rare on Öland, the answer was that the deer enter the gardens during the winter, when the snow cover is higher than the walls and eat the trees, while at the same time the hares bark them.

Pits for catching hares were made in the same way as for wolves in the winter; the pit is made in the snow, cabbage is placed on the brink and the trap is half covered with snow; a great many hares are caught in this fashion.

Hawthorn, which it is permitted here in some places to cut down, has a very hard and tough wood, which is used for mill wheels and cribs.

We stayed at Pesnäs during the night.

June 13 [p. 118]

We left in the morning in the direction of Gaxa, where the main road ends.

Grass worms were seen moving in flocks in the meadows, destroying the crop; they were of another type than those which did so much harm to the grass in Uppland and Dalarna recently; each grub (*Eruca*) was about a finger long, had 16 feet, and the entire body was covered with yellow hair, the head was grey with grey hair, and on each side along the body was a broad bluish line here and there intersected with black, and between them a narrow yellow line. The back is black with four longitudinal lines, interrupted at the indentations. We threw it to the hens, but they did not dare to try this new food.

The road here was newly and carefully made, covered with sand and separated from the pastures with ditches. A big group of country girls, under the command of the local sheriff, had been working on it for the last four days and made a road that could hardly be better. They used scouring clay from a hill where there had formerly been a grinding mill, and I cannot think of a more suitable material for roads.

The journey proceeded through a young oak wood, with hazel and beautiful grass at Långrum village.

In the ditches we saw the black soil, 1-1½ quarter-ells [15-22 cm] deep and under it white sand from the sea.

Young oaks, in 20 pairs, were planted along the two sides of the road like an alley; they were protected from cattle by poles, as high as a man and bound together; if the cattle and wild horses did not destroy the young oaks during the winter there would be many more trees here.

Sandy hill ridges were seen in a few places cutting transversely across the land, sloping down to the eastern shore where they ended as banks in the water.

We reached Gaxa at noon and decided to wait there for the mail boat that would take us to Gotland.

White sand, thrown up by the sea, lay on the shore in large quantities, the same as we had seen along the road today, which was the cause of the sandy fields and the short grass. At the water edge the seaweed was packed together, and sand was thrown up on it forming banks and islets. This is surely the way in which Nature has gradually built all the land here.

Seaweed [*Fucus vesiculosus* L.] lay on all the shore, a good fertilizer for sandy fields, but not for the black soil; it produces a red colour, which fastens itself above the *Conferva* [i.e. stains this].

"Kräkel" is the Ölanders' name for the seaweed that Botanists call *Fucus dichotomus ramosissimus teres uniformis fastigiatus* [*Furcellaria fastigiata* (L.) Lamour.], which we have never seen before. It is not thicker than a thread, rounded and densely branched and bifurcated, and all branches were of equal length, which made it look as if it were cut off at the top.

Potentilla [*P. anserina* L.] grew directly on the sand, and *Senecio minor vulgaris* [*S. vulgaris* L.] on the rotting seaweed.

Phalaena seticornis spirilinguis, alis superioribus fuscis: linea punctisque duobus rubris, inferioribus rubris [*Hipocrita jacobeae* (L.), cinnabar moth], a beautiful butterfly, flew everywhere on the shore; its larva (*Eruca*) grazes on the above-mentioned *Senecio.* The body was black, the antennae bristlelike, it had two little spiralled tongues; the upper wing had a red longitudinal line along the front edge; at the point and the back edge there was a red spot; the back wing was bright red but black along the front margin.

Cimex oblongus niger, elytris lineolis minutissimus sulcatis, alis pone flavo maculatis [*Acanthia saltatoria* (L.)] jumped in damp places on the shore and had a bad smell. It was completely black; a few dark yellow dots on the wings above the tail and a number of black ridges on the wings could be discerned with great difficulty; the horns consisted of four even joints without hairs at the end.

Cimex oblongus niger, elytris cinereis antice nigris, alis pone albis [*Trapezonotus arenarius* (L.)] jumped in the sand; it was completely black except for the wings; the cross on the wings was black in the front, white in the back and had grey sides with small black dots, darker on the back.

Cicindela viridi-aenea punctis latis excavatis [*Elaphrus riparius* (L.)], a little bright green insect with a copper hue, jumped rapidly on the shore. Head, breast, abdomen, thighs and feet (but not legs) glittered like copper or gold; the neck was thin and the dark eyes protruded from the skull; the black antennae had 11 joints; the wing-cases were green with a copper tinge and had large hollowed-out dots with a smaller elevated dot inside them; the larger dots were joined with an elevated line.

Palings around the farms, which are commonly used over the whole island (*vid.* June 2), were here nothing but unhewn wood placed on the sides, serving to enclose the farm and give protection against the violence of storms.

Vicia segetum [*V. cracca* L.], *Ophrys major* [*Listera ovata* (L.) R. Br.], *Ribes* [*R. alpinum* L.] and *Frangula* [*F. alnus* Miller] grew in the meadow between the church and Gaxa.

Orchis muscam referens [*Ophrys insectifera* L.] grew abundantly in the same meadow, and we never saw more of it anywhere else; the sepals were separated from the petals so that the former were green but the latter narrow and brown. The stamens were each attached to a little elevated smooth dot.

Aira foliis pubescentibus, panicula contracta, flosculo hermaphrodito mutico, masculo, arista uncinata calice breviore or the handsome grass *Gramen lanatum Dalechampi* [*Holcus lanatus* L.], which is so common in Holland but has not been seen in Sweden before, grew in several places around Gaxa.

The day was calm and nice; the clouds formed a rounded *lantborg* at the horizon after the sunset.

We remained in Gaxa during the night to wait for the ship.

June 14 [p. 122]

This Sunday, which was the 4th after Trinity, we celebrated in Högby church where we saw a procession moving towards the altar as we entered.

The procession was thus: first came nine girls, then three women carrying children, then three women who were to be churched, then nine pairs of farm wives, then half a dozen farm-hands, and they all proceeded with firm steps, the girls to the south of the altar and the men to the north, those who were carrying the children to the altar rails, then back again, and the three mothers knelt in front of the rails and received the benediction. After this was all done, the girls went up to the altar, two at a time, courtseyed, offered a coin and went away. All were carrying leafy twigs in their hands, instead of fans, except those who were carrying the children. Their garments were as follows: on the head they had a cloth tied in the neck and smoothed with a glass stone; the breast was covered with a linen cloth ironed in the same way with glass stone, and on top of this a sleeveless vest sewn to the skirt, and a belt. Some neckerchiefs were embroidered with silk, others were edged with lace. The married women had shawls of Västergötland cloth, such as is used for tablecloths, 1½ quarter-ells [22 cm, 9 inches] broad, and with fringes like "kantilier"; girls did not wear these, but instead the ribbons in their hair hung down below the headcloth. Those who carried the children, and the mothers, had hip-long sleeveless coats of heavy black valmar cloth and the sleeve openings were edged with a piece of cloth shaped like a horseshoe; the collar was stiff and arranged horizontally around the neck; the aprons were blue yellow or green and quite small. The farm-hands had shoes with narrow heels and heavy soles, interleaved with birch bark. The trousers were of valmar cloth, usually grey, and as wide as sailors' trousers. The blouses were short. The hat was, as in the rest of the country, without a band and the brim was horizontal at the base and upcurved in the edge.

A ship model hung in the church like a lamp, as we had seen in churches over almost the whole country.

There was a runic stone in the churchyard, outside the porch, which was remarkable in that the letters were the wrong way and went from right to left; only one side was reasonably legible:

·····ᛆ×ᛚᛆᛐᛌ×ᚢ\ᚦᛆᛁᛁ×ᛉᛁᛉ:ᚢ,ᛁᚱ:ᛁᚱᚢᛁᛒ:ᛉᚢ··

Linnaeus (reading from the right to the left):

. . . uk : biuri : ri-u : kik × *iiaþ-u* × *stal* × *a . . .*

Öl 55: *. . . auk : biurn : risu : . . .*

"*. . .* and Björn raised . . ."

"Öland horse", as one calls a very small horse in Sweden, is however far from common on Öland without being very rare; all the horses on Öland are small and good runners with steady feet which do not slip or stumble even on the *alvar*-land, but they are far bigger than the so-called Öland horses; when I asked the reason for this, I was told that, when the inhabitants were forbidden to sell them, they did not care about breeding the small horses which are good for sale but not for farm work.

There were many bees on the island formerly, but the last two hard winters have devastated most hives and there are few left.

The honey was rather white, due to lack of heather, from which the bees in Småland draw their brown honey.

The fences here were constructed of timber, as in the rest of Sweden, indicating that one was in Öland's northern and wood-rich district. Service trees [*Sorbus*] grew both large and abundantly.

There was cummin on the ditch banks, much sawwort in the meadows and *Papaver erraticum* [*P. dubium* L.] in the fields.

The farmers use wooden soles, tied to the foot like skates, when they work in the quarries.

This night, after a sunny warm and quiet day, we again stayed in Gaxa.

June 15 [p. 125]

At 4 o'clock in the morning we were on horseback heading for Blåkulla. The way ran west of Gaxa, from the eastern to the western shore, ½ a mile long, through bushes, hazel and oak woods, mixed with some spruce. The weather was pleasant, the sky clear, and larks, swallows, cuckoos and the recently fledged starlings were calling around us. When we arrived at the western shore the boat was on land, the oars gone, the boatmen stubborn and everything in disorder; while this was arranged, a gale blew up which delayed our voyage to Blåkulla until the afternoon, but we spent the time wandering about observing things to illustrate the natural history of our country.

The sea was making ridges up to 30 ells [18 m] broad and a few fathoms high of rather large stones which are here and there intermingled with seaweed on the shore, situated outside the *lantborg.* Thus the land now, as formerly, increases in width and forms new *lantborgen.*

Spiders had made innumerable webs among the gravel on the shore, constructing them horizontally along the earth surface and themselves living below them. They are of *specie Araneae thorace testaceo, abdomine ovato nigro-aeneo apice sub luteo bicorni* [*Agelena labyrinthica* (L.)] ; the entire body was covered with hair; so that when seen against the sun the colour of the abdomen had a copper hue, the processes at the tail were small.

Incredible amounts of *Asclepias* [*Vincetoxicum hirundinaria* Medicus] grew on the shore and always close to the rocks; the cattle did not touch it, but people said that they will eat of it in the autumn when it is withered and there is nothing else to have.

"Käringetänder" (*Euonymus*) [*E. europaeus* L.] is the name of a rare tree which grows only here and in Skåne and is used for hedges in gardens but here was 3 fathoms [18 feet, about 6 m] tall and as broad as a man.

The birches had a multitude of caterpillars, which had devoured the leaves and surrounded themselves with a woven net; they were black and covered with long white hair; on the back they had two tufts of yellow-brown hair close to the neck; the rear and middle feet were red.

Ossea [*Cornus sanguinea* L. = *Swida sanguinea* (L.) Opiz] had another kind of caterpillar, which was as large and broad as a finger, covered with hair, yellowish with eight broad transverse black bands without hair and many fine white lines coming together at an acute angle; it had 16 feet.

The scabious (*Scabiosa*) [probably *Knautia arvensis* (L.) Coulter] also had its own caterpillars with a hirsute head and 16 feet; the rest of the body was hairless and they were green except for a yellow longitudinal line on the sides.

The smaller variety of nettle [*Urtica*] was also almost entirely devoured by its own caterpillar, which is the well known *Papilio urticaria vulgatissima Raj.* [*Aglais urticae* (L.), small tortoiseshell butterfly] .

Aranea abdomine ovato nigro, annulo ovali dorsali albo [*Aranea corollata* L., *Lithyphantes albomaculatus* (de Geer)] , a spider with a white line which was two-cleft on each side and enclosing an oval, with several white teeth on the inside margin, was found among the bushes.

A mite, an *Acarus* [*Ixodes reduvius* (L.)] with an oval spot on the back towards the breast, surrounded by stripes, ran on the ground.

Cantharis aeneo-viridis, elytris apice rubris [*Malachius bipustulatus* (L.)] , a delightful little insect, sitting on the leaves of the plants, had black antennae and feet; when it was squeezed a blood red vesicle appeared on each side of the abdomen, also the breast, which is very remarkable.

The linden trees had little red protrusions on their leaves which might be called gall apples in a sense.

Cochlea vulgaris testa variegata Petiv: mus 5.n.14 [*Helix nemoralis* L.] of several colours was seen in the groves.

The way went to the Horn estate, which lies close to the sea. The beautiful meadows were full of lime trees and other beautiful deciduous trees.

The most characteristic plants in the meadow were *Vicia multiflora C.B.* [*V. cracca* L.], *Lathyrus pratensis Riv.* [*L. pratensis* L.], *Dentaria* [*D. bulbifera* L.], *Milium* [*M. effusum* L.], *Astragalus* [*A. glycyphyllus* L.], *Scabiosa pratensis hirsuta* [*Knautia arvensis* (L.) Coulter] *Galium palustre album* [*G. palustre* L.], *Veronica supina & vulgatissima* [*V.* 8 *officinalis* of Index = *V. officinalis* L.], *Iris* [*I. pseudacorus* L.], *Jacobea vulgaris laciniata* [*Senecio jacobaea* L.], *Cichorium* [*C. intybus* L.], *Thlaspi arvense siliquis latis C.B.* [*Th. arvense* L.] and *Filipendula vulgaris C.B.* [*F. vulgaris* Moench].

Asclepias [*Vincetoxicum hirundinaria* Medicus] grew in the driest and stoniest places; *Melica 1* [*M. ciliata* L.] grew where the soil was most humid with broad leaves a quarter or a ½ ell [15-30 cm] long; *Alopecurus erectus* [*A. pratensis* L.] grew in marshy places, where all other grasses are short; *Asperugo* [*A. procumbens* L.] abounded and was very tall, as in Germany, and *Papaver erraticum* [*P. dubium* L.] adorned the corn fields together with the charlock.

Serratula, quae Carduus in avena [*Serratula* 662 *arvensis* of Index = *Cirsium arvense* (L.) Scop.] was common in the fields, since the farmers leave it at liberty to flourish and go to seed when the fields lie fallow.

We had a quick look at the *alvar*-land, where the soil was tough like clay from the little rain that had fallen recently. On the *alvar*-land *Sedum minus teretifolium C.B.* [*S. album* L.], *Veronica spicata minor C.B.* [*V. spicata* L.], *Artemisia, quae Arbotanum campestre C.B.* [*A. campestris* L.] grew in large numbers.

A little lake, a ½ mile long and a ¼ mile broad, was seen north-west of Horns ladugård.

The journey to Blåkulla was undertaken immediately after noon, after a hearty meal, when the wind and the sea had calmed down. We hurried to the shore and steered the boat towards Blåkulla, a rock two miles away, all blue and like a hemisphere rising out of the water, and it seemed as if it were running away from our hard-working boatmen. While they were telling us that it should not be called Blåkulla, but rather Känningen or Jungfrun [The Maiden], since otherwise there would be storm and mortal danger, a northern wind started to blow and to heave up the waves, and a violent storm shook the boat; we all had to work hard with danger to our lives; and we finally arrived, dead tired, after almost being crushed against the rocks.

Blåkulla is a little island, situated between the northern point of Öland and Småland, which old hags and fairy tales dedicate to Pluto, and not to Neptune, who however seems to be more eager to protect it from nicknames; the people say that all the witches have to go here (a troublesome journey indeed) each Maundy Thursday, but anyone who has visited the place once is unlikely to return; if any place in the world looks dreadful, this is it, and the description will be appropriately short. It has a rock wall and within this a low deciduous forest of oak, birch, etc., then rock again, more low woodland, still higher rocks with the highest in the centre; the rock consists of a red spar-like stone, overgrown with a pitch-black lichen, *Lichenoides 81 Dil.* (p. 54) [see June 2 entry; *Parmelia stygia* (L.) Ach. but the lichen intended is probably *P. dentata*], which covers the rocks and colours them black. Among them *Lichen cinereus arboreus marginibus pilosis major Vail.* [*Anaptychia ciliaris* (L.) Körb., but probably a wrong identification] grows directly out of the rock; these two, together with the *meteora* [light effect, haze] rising out of the sea, cause the blue colour of Blåkulla when it is seen at a distance. With much labour we managed to climb the high steep rock between the large boulders which look as if they had been thrown there. The deciduous trees grew in groves like little garden terraces and were hardly two fathoms tall [12 feet, 3.6 m], although the oak trunks were often as thick as a man; we also saw oaks growing on the naked rock, with recumbent trunks and branches, as if they were creeping on the ground. The thicket was so dense that a man could hardly penetrate it, and, what was remarkable, the shrubs were also entwined with *Hedera arborea* [*H. helix* L.] which grows all over the place, like peas and bindweed. We were eager not to leave the island until we had reached its centre, where we had a good view over it all: the wild sea to the north and the south, Öland in the east with several churches and Borgholm Castle and Småland in the west with a little island call Förön; there was less wood on the northern part of the island, the land was more level and the rocks were battered by the waves. To the south there was a cave, like a chamber, in the rock. On the top of the mountain were dried-out marshes. There were deep crevices and channels, hewn out in the rock and ground by the sea, and even in the highest rocks there were marks, like waves, as a sign that the sea had raged here too in former times. Smooth boulders were thrown up on the land, as far from the water edge as a man can throw a stone, and piles of such stones were seen on top of the highest rocks; these stones were, however, covered with *Lichene crustaceo leproso,* a sign, that it was not yesterday they were touched by the waves of the sea. Here were no signs of human activity, only a little pile of loose stones on a rock, undoubtedly erected by some seaman detained by contrary winds. Here was no living creature except a wild he-goat jumping on the rocks together with his mate, and some black scoters [*Melanitta fusca* (L.), the velvet scoter] around the shores. A dead bull-head fish [*Onocottus quadricornis* (L.)] without horns or warts was thrown up on the shore with its sharp needle-like rays on the back of its head. Major Hammarskiöld was said to have bought this island at a cost of 4 *Daler S:Mt.* The sailors told us that last year a lieutenant had sent goats here together with a girl to tend them during the summer; storm or some other adversities intervened and the herdsmaid did not receive any food for the whole time. Thus she had to subsist in great distress on grass and raw goat meat, because she had no fire, until a ship arrived and rescued her from Blåkulla.

I will enumerate the plants on Blåkulla, so that the botanists will have a reliable flora of the place.

**Veronica mas* [*V. officinalis* L.]
Pseudochamaedrys Till. [*V. chamaedrys* L.]
Anthoxanthon [*Anthoxanthum odoratum* L.]
Valeriana 1 (copiose) [*V. officinalis* L.]
†*Eriophorum culmo tereti* [*E. vaginatum* L.]
Melica 1 [*M. nutans* L.]
Aira 1 [*Deschampsia cespitosa* (L.) Beauv.]
2. Alpina [*Deschampsia flexuosa* (L.) Trin.]
Spica lavendula [*Molinia caerulea* (L.) Moench]
Festuca Sheep's fescue [*F. ovina* L.]
Secale 2. maritima [*Elymus arenarius* L.]
Galium luteum [*G. verum* L.]
Cruciata aqu. [*G. palustre* L.]
Aparine vulgaris [*Galium aparine* L.]
Parisiensis [*Galium uliginosum* L.]
Asperula odorata [*Galium odoratum* (L.) Scop.],
Plantago lanceolata [*P. lanceolata* L.]
Myosotis [*M.* species]
Xylosteum [*Lonicera xylosteum* L.]
Ribes [*R. alpinum* L.]
Hedera [*H. helix* L.]
Asclepias [*Vincetoxicum hirundinaria* Medicus]
Angelica sativa [*A. archangelica* L. subsp. *litoralis* (Fries) Thell.]

* Larger and more luxurious than in any other place.

† *Eriophorum culmo tereti, spica ovata* in the crevices in the highest rocks.

Thysselinum [*Peucedanum palustre* (L.) Moench]
Opulus [*Viburnum opulus* L.]
Cepa Alwarlök [*Allium schoenoprasum* L.]
Convallaria 1 [*C. majalis* L.]
Unifolium [*Maianthemum bifolium* (L.) Schmidt]
Polygonatum [*Polygonatum odoratum* (Miller) Druce]
Juncus culmo stricto, cap. lat. [*J.conglomeratus* L.]
Capit. psyllii [*Luzula multiflora* (Retz.) Lej.]
Rumex folio crispo [*R. crispus* L.]
Herba Britannica [*R.* species]
Acetosa pratens. [*R. acetosa* L.]
Acetosa lanceolat. [*R. acetosella* L.]
Epilobium latifol. [*E. angustifolium* L.]
Vaccinia rubr. [*Vaccinium vitis-idaea* L.]
nigr. [*Vaccinium myrtillus* L.]
Erica vulg. (altissima) [*Calluna vulgaris* (L.) Hull]
Saxifraga alba [*S. granulata* L.]
Alsine media [*Stellaria media* (L.) Vill.]
Silene viscaria [*Lychnis viscaria* L.]
Mont. hirsut. [*Silene nutans* L.]
Cerastium 1 [*C. holosteoides* Fries]
Sedum album [*S. album* L.]
Acre [*S. acre* L.]
Telephium [*S. telephium* L.]
Prunus sylv. [*P. spinosa* L.]
Cotoneaster [*C. integerrimus* Medicus]
Sorbus [*S. aucuparia* L.]
Oxyacantha [*Crataegus oxyacantha* L.]
Rosa sylv. [*R. canina* L.]
Rubus fr. caesio [*R. caesius* L.]
Fragaria [*F. vesca* L.]
Geum luteum [*G. urbanum* L.]
Tilia [*T. cordata* Miller]
Ranunculus Flammula [*R. flammula* L.]
repens hirsutus [*R. repens* L.]
Ficaria (copisissime) [*R. ficaria* L.]
Anemone nemorosa [*A. nemorosa* L.]
Hepatica [*H. nobilis* Miller]
Linnaea [*L. borealis* L.]
Melampyrum lut. [*M. pratense* L.]
spica tetrag. [*M. cristatum* L.]
Dentaria [*D. bulbifera* L.]
Geranium Uniflor. [*G. sanguineum* L.]
Lathyrus sylv. major [*L. sylvestris* L.]
Sylv. minor [*L. pratensis* L.]
Cicer Cracca minor [*Vicia hirsuta* L.]
‡*Vicia sepium* [*V. sepium* L.]
Cracca seget. [*Vicia cracca* L.]
Lotus 1 [*L. corniculatus* L.]
Hypericum caule ancip. [*H. perforatum* L.]
Tanacetum [*Chrysanthemum vulgare* (L.) Bernh.]
Solidago [*S. virgaurea* L.]
Millefolium [*Achillea millefolium* L.]
Orchis alba [*Platanthera bifolia* (L.) Rich.]
long. calcarib. [*Gymnadenia conopsea* (L.) R.Br.]
Carex nigroluteus [*C. acuta* L.]
Betula [*B. verrucosa* Ehrh.]
Pinus [*P. sylvestris* L.]
Abies [*Picea abies* (L.) Karsten]
Quercus [*Q. robur* L.]
Salix fol. myrti. [*S. repens* L.]
Populus [*P. tremula* L.]
Juniperus [*J. communis* L.]
Fraxinus [*F. excelsior* L.]
Pteris [*Pteridium aquilinum* (L.) Kuhn]
Polytrichum vulg. [*P. commune* Hedwig]
Mnium caule simpl. [*Philonotis fontana* (Hedwig) Bridel]
Bryum bulbiforme [*Funaria hygrometrica* Hedwig]
Lichen rhangiferin. [*Cladonia rangiferina* (L.) Weber]
crustac. lepros. [?]
§*Lichenoides Dil.81* [*Parmelia stygia* (L.) Ach.]

‡ *Vicia sepium* was so large, that each *foliolum* was bigger than the end joint of the thumb.

§ *Lichenoides Dill 81.* If this moss should ever be needed for dyeing bright red in the dye-works it ought to be collected here, since it is rare in the rest of Sweden. See Borgholm, June 2.

We left Blåkulla in the late afternoon, wind and waves propelled our boat; Blåkulla lay between the sun and our little ship; a gale started to move the waves as we came onshore at 10.30 p.m.; we rode back to Gaxa immediately.

June 16 [p. 132]

Today we had to wait in Gaxa for a better wind which would enable the boat to sail back to Gotland whence it had now arrived; in the meantime we investigated the economy of the place.

The lime bedrock yields a densely growing grass which is however so short that a whole acre will often yield less than an armful of hay. We asked whether ditching would improve the situation, but the unanimous answer was that it made no change.

The gravel does not protect against drought since the soil layer on the bedrock is scarcely a quarter-ell [15 cm] deep.

The village community owns the fields and the meadows here, which are all enclosed.

Rye was common here in the northern parts; barley, however, was quite rare because of the sandy soil.

Oats are hardly ever grown except a little at Mörbylånga. Wild oats grow in the northern district only.

No remedy is known against charlock, which is here called "åkerkåhl"; it seeds before the crops ripen.

Kale is seldom cultivated here; we were surprised that the farmers could manage without this useful vegetable.

Beans are never grown although the climate is suitable.

The inhabitants are not accustomed to flax culture, and there is little hemp to be seen, hence the farmers are poor in linen clothes since they have to buy their linen from Västergötland.

The economy for the inhabitants of the northern district is rather limited. They have no hay, grain, cattle, butter or cheese to sell but have to earn money by lime-burning or quarrying; no burnbeating is done, no potash is derived from the oaks, and tar from roots and stubs is made in Böda parish only. Anyone, including those who live on the western shore, is allowed to fish herring and cod in the Baltic. Goats are not present and should not be introduced here. One sees few beehives here in the north. Tile stoves are not known on Öland, and yet firewood has to be bought, and in the northern part, where there is wood, the trees must not be cut until the Governor or the Court has allocated wood to the villages and the Crown Forester has given special permits to the farmers to cut the trees which are marked out by the wood-warden.

Thorn can be cut by anyone. By thorn is meant all bushes which carry thorns, or do not grow to tall trees, such as hawthorn, blackthorn, buckthorn, dog rose and bird cherry.

It is prohibited by law to cut branches from oaks here, and the penalty is 2 *Dal. S:mt.;* therefore the oaks are not as mutilated as in Östergötland. In spite of the existence of several nursery gardens the planted oaks do not thrive well; they are most often small and spindly, but when they grow among the hazel, they grow better. No swine are sent here at acorn time, either from Småland or from other places.

Lime burning was not considered to damage the woods since only windfall is used for fuel, but it was admitted that the farmers were sometimes allotted growing trees for the lime-kilns.

We asked frequently for eiderdown, but it was not collected here.

The hay bins for the deer are enclosed with fences up to 20 paces from the barn, in order to prevent the wild horses from feeding on the hay. The deer jump easily over the fences, while the horses do not.

In the evening we heard the nightingales sing and we saw that little four-footed animal, the bat, playing in the air like a bird.

The wind was all day from the south-east, with a furious storm, and no ship could depart, so we spent another day in Gaxa.

June 17 [p. 134]

In the morning the captain called us on board, but the same strong sidewind continued with equal force, and we were warned against embarking on the fragile vessel, which had already been condemned and was due to be repaired immediately after its return to Gotland; there was also the disadvantage that the ship could not tack without using oars; we hesitated in front of this obvious danger and let the captain leave without us. For the sake of travellers as well as for the Gotlanders themselves, I must mention how essential it is to have a safe ship to ply on this route, which is the only connection with Gotland.

Thus, we left Gaxa and Högby church towards Böda and so further to the east side of Öland. The road passed through woods of oak and hazel, alternating with spruce.

There were a few burial mounds along the road. In one place there was a ring of seven stones, each resting on three smaller stones.

The fences were like in the rest of Sweden, but taller and with less space between the poles.

The halters which people here had for their horses were very strange; they consisted of thin rods on each side of the horse's head, fastened with a strap over the head, behind the ears; another strap was attached to the centre of the rod, and a rope was attached to the lower end of one rod and threaded through the lower end of the other.

We wanted to see the islet outside Böda which forms a delightful harbour for a few vessels, and thus we rode out on a narrow promontory from where we noticed that the islet was entirely surrounded by water; since we had been informed that it dries up here in the summer, we tried to pass on horseback at the narrowest place, but when we came so far out that the horses were forced to swim, they did not manage and we had to return; one member of the company however was forced to jump into the water since his horse was entangled in the seaweed and he arrived at the islet only after much labour. On our way back we noticed that there was a white streak on the water at a place in front of the islet and we saw horses' footprints under water, in the white sand, and on this track we arrived easily at the islet. There was nothing very interesting there except *Carduus lanceolatus latifolius vulgaris* [*C.* 654 *lanceolatus* of Index = *Cirsium vulgare* (Savi) Tenore], which grew on the shores. At a corner of the island was a lime kiln with some broken limestone. In the limestone were several petrifactions or "Ölandispikar", which are called "darter", *entrochi* with many rings, round and uneven blocks of pyrites, and "crystal apples".

I call "crystal apples" the globular stones found in the limestone, which are as big as an apple, look like *haematites* when they are broken, and consist of light transparent spar crystals, which are joined in the centre, sometimes with a little hole in the middle where their triangular processes show. These crystal apples are quite common on Öland; I have also collected them at the Osmund mountain in Dalarna; see *K. svenska VetenskAkad. Handl.,* 1740: Tab. 2, Fig. 18.

Böda church, which stands 1 mile from Gaxa, was visited. We asked about runic stones which might have been there, but saw nothing but a few stones with black letters. The floors in the houses were strewn with *Ulmaria* [*Filipendula ulmaria* (L.) Maxim.] and smelled so strong that one could hardly bear it. The vicar told us that the inhabitants were very worried about our journey, that our trip to Blåkulla was known and that they took us for spies.

The meadows between the church and the sea were said to have an abundance of rare plants, but there was nothing exceptional there, only *Orchis muscam referens* [*Ophrys insectifera* L.], *Orchis militaris minima* [*O. ustulata* L.], *Melampyrum arvense, Trifolium lupulinum* [*T.* 617 *lupulinum* of Index = *T. agrarium* L.], *Aira: gramen lanatum Dalech.* [*Holcus lanatus* L.].

Wild chervil [*Anthriscus sylvestris* (L.) Hoffm.] grew in the hop-gardens, as high as a man can reach with his outstretched arm.

We saw here "Äggskalar", a small kind of bread, which is as well known in Sweden as it is rare abroad; it is as big as a cake, and shaped like a watchglass. It is made from egg and flour without any additions; the cakes are cut out from the dough with a drinking glass and are coated with melted butter and sprinkled with sugar; this clogs the pores on the upper side so that the centre of the little cake rises in the hot oven.

From Böda we proceeded northwards, since we wanted to see the northern point of Öland. The road passed through a big forest to Sjötorp.

The wood consisted of fir, spruce and juniper; the fir trees, which are good for timber, lift their branches towards the sun; the spruce twigs droop, and the junipers are bushy. Reindeer moss covered the higher spots. Whortleberries, cowberries and heather coloured the ground green. My flower [*Linnaea borealis* L.!] grew in the deep forest with its pairs of drooping flowers, and indicated that the place has not been burnt within living memory. *Trientalis* [*T. europaea* L.] grew on the slopes with flowers bent down to the earth, since it had been raining all day. The marsh rosemary [*Andromeda polifolia* L.] was in flower in the marshy places and *Sparganium* grew in small pools, and also bracken (*Pteris*) [*Pteridium aquilinum* (L.) Kuhn] here and there.

Sjötorp, a farm 1 mile from Böda, was the end point of today's journey over sand to a beautiful black soil; the farm was enclosed within high fences as protection against the deer. Here we were allowed to shelter from rain and darkness.

June 18 [p. 138]

From Sjötorp we proceeded to the next hamlet, Grankulla. As soon as we came here, we saw that the whole region between the sea and the hamlet was full of sandhills.

Drifting sand was thrown up from the sea by a strong southerly wind, flew northwards, covered the district and was not stopped until it reached the stillness of the woods, where the storm's violence lost its force; we saw piles of sand, like snow drifts, at the edge of the forest, burying the fir trees, so that often only a third of the crown emerged from the sand on the inner side of the drifts. Thus the trees were slowly suffocated because no rain and moisture passes through the sand to nourish the trees. The outermost trees perish, and the sand drives more and more into the wood each year. It was nice to run on these sand drifts and to botanize between the treetops. The sand was not at all like the drifting sand in Skåne, it was much coarser but all white; it consisted of clear quartz and a little reddish spar, but was too uneven to be used for scouring. The shore side of the sand drifts sloped moderately with waves and ridges as on the bottom of the sea, but the land side was almost too steep to be climbed, and about 4 fathoms [24 feet, 7.2 m] high.

"Sandhavre", *Arundo foliorum lateribus convolutis: acumine pungente* [*A.* 102 *arenaria* of Index = *Ammophila arenaria* (L.) Link, marram grass] grew in tussocks on this sand-dune and prevented the sand from blowing away; we tried to dig up the roots but they went so deep that we could not find the bottom and were as thick as binding twine with innumerable ramifications and at the top there was a brush of old withered and stiff leaves. This sand-grass is the same one as the Dutch plant on their dunes to retain the sand, and it is also recommended for use in Skåne.

The sand sedge, *Carex spica composita, spiculis androgynis, inferioribus remotioribus foliolo longiori instructis* [*C.* 749 *arenaria* of Index = *C. arenaria* L.], a grass which has not been discovered in Sweden before, grows under the sand with creeping roots some fathoms long, puts forth little tufts of stalks and leaves evenly spaced at a quarter-ell [15 cm] apart, as if it were planted along a string. Nature herself teaches us to use this grass for retaining the sand, although no one has thought about it before. This grass and the previous one were the only plants that can survive on the sand-drifts.

On the area between the sea and the sandhills, where the sand-drifts had passed, nothing was left but a few thin firs; from these Böda church could be seen in the south.

Iberis foliis sinuatis, caule nudo simplici [*I.* 536 *minima* of Index = *Teesdalia nudicaulis* (L.) R.Br.], a plant which I have seen in only a few places in Skåne and which has not yet been included in the Swedish flora, grew in the forest, on the sand-drifts. *Radix annua, minima. Folia radicalia prima ovata, integra; reliqua lyrata s. Bursae pastoris simillima. Caules plures, digitales & spithamei, nudi aut unico foliolo instructi. Racemus terminatrix. Flores albi: petalis duobus exterioribus duplo aut triplo majoribus. Stamina sex. Capsula subrotunda: loculamentis singulis duo semina includentibus, latere inferiore gibbis, apice emarginata.*

We left the sandhills and proceeded westwards to the bay. *Jasione* [*J. montana* L.] and *Spergula* [*S. arvensis* L.] grew in the sandy field, which had to lie fallow for four or five years after one year's cultivation. On the road we saw tar being distilled and lime was burnt at the bay.

The tar was distilled thus: a pit was dug in the ground, shaped like an inverted cone; the sides were built up with stone and clay; it was 1 fathom [6 feet, 1.8 m] deep and the top diameter was 1½ fathoms [9 feet, 2.7 m]. The pit is filled above the brim with tarwood, and the construction is caulked with peat and earth, so that the flame emerges from the top only, about a quarter high. On both sides were little walls of rubble, 1 ell [2 feet, 60 cm] from the pit. A man was continuously tending the pit and tightening the side facing the wind to prevent the opening from becoming too wide. At the bottom of the pit was a closed barrel with an outlet for the tar at the bottom. One pit was said to yield half a barrel of tar daily and would burn for about eight days. After burning there was also a quantity of charcoal.

The lime kilns, of which there were many on the bay, were with their walls 5 ells [10 feet, 3 m] and 4 ells [8 feet, 2.4 m] broad, of the same shape as a roasting furnace, made of stone and covered on the outside with earth. The limestone was laid into the kiln, and two parallel furnaces were constructed from it, 1 fathom long [6 feet, 1.8 m] long, 1 ell [2 feet, 60 cm] broad and 1½ ells [3 feet, 90 cm] high. Both these little furnaces opened at the base of the kiln. When these had been made, the kiln was filled with limestone and the furnaces were filled with spruce wood; a fire was then lit which burnt for two days, and fresh fuel was added continuously; about 60 loads of wood are needed for the two days. It is easy to see when the lime is sufficiently burnt because it turns yellow and does not burn as readily any more. The burned lime is removed and the stones which are insufficiently burnt are separated from it. They are easily recognized since they are heavier than the rest and have a brown discolouration. Today the workers were carrying water from the sea to slake the lime. This kind of kiln yields 20 loads of slaked lime, with 12 barrels on each load; the ships which carry the lime to Stockholm take 20-30 loads, and the price is 7-9 *Daler Km:t* a load. The farmers here, whose fields are mostly sandy and their meadows dry and poor, can only by the sale of lime earn money for taxes and subsistence.

The limestone for quarrying was red and green and lay ½ an ell [1 foot, 30 cm] under the earth surface; the stone was identical with the one that is quarried in the *alvar*-land for floor stones, but thicker and more uneven.

"Darter" or *Helmintolithus nautili recti* [*orthoceratites*] were present in quantity here, especially in a red stone with blue streaks. Mostly they were of reddish, often a dirty red colour, on the outside, like blood-stone; some of them were hollow between the *dissepimenta* and filled with white spar crystals; most of them were of the same shape as salt from the sea but others had more facets. We searched in vain over all Öland for the snailshell on the beach, of which the "darter" found here in all rocks are obviously petrifactions, since one often finds "darter" hollow so that *crusta* and *dissepimenta* are all empty.

Eschara membranacea covered most of the limestone in the quarries close to the sea; these were now filled with water, but they dry up in the autumn and this is the time when they are quarried.

Bufo, a big toad, was seen among the limestone boulders; it was black with coarse warts; the back feet were webbed like those of a goose with six fingers, the innermost being the smallest, and the rest of increasing length except the outer which was slightly shorter. The fore feet had four toes of almost equal length. There was an oblong slightly elevated vesicle at the ears. The skin under the chin was yellow. The abdomen was broad, pale and covered with small dark spots as were the thighs, which were still darker [=*Bufo bufo* (L.)].

The stock doves [*Columba oenas* L.] flew in flocks in the wood and among them were also ring doves [*Columba palumbus* L.]; the latter were called "skjututs" here because of their call.

The journey continued along the eastern shore to the west cape, which is the most northerly point of Öland. Here we saw nothing new; only a few wrecked ships adorned the shore, since in this place they are an adornment.

"Ship-wrecks" is my name for the decorations taken from the unlucky ships that are wrecked on these coasts. One can see them above gates and outhouses, figures of seahorses, whales, lions and saints, which have obviously been brought here before the Dykerisocietet [Salvage Company] was formed. No spectacle could be more unwelcome for those who travel to the northern part of Öland and from there trust themselves to the treacherous sea.

The outermost point of the west cape forms a little islet in the east. On this point is a level place as big as a market place, which is under sea-water during the winter. Here grows *Plantago foliis semicylindraceis integerrimis* [*P. maritima* L.] and a strange *Lotus.*

Lotus leguminibus solitariis membranaceo-quadrangularibus foliolis floralibus lanceolatis [*L.* 610 *marina* of Index = *Lotus maritimus* L.], a plant which we have never seen before, neither here nor in Sweden, occurred here in some profusion; it was similar to *Lotus tetragonolobus,* but the roots were perennial and the flowers yellow. The stem, which arose from a deep root, was only a few inches tall, with a few tripartite stalked leaves of which the leaflets were glossy, fleshy, ovate and stalkless. The stipules were ovate, entire, longer than the leaf-stalks and almost as big as the leaflets. The flower-stalk, longer than the stem, emerged from between the stipules and carried one terminal flower as large as a bean flower. Immediately below the flower sat a tripartite stalkless leaf with lanceolate leaflets. The calyx was deeply cut and brownish on the upper side. Of the petals the standard was pale yellow with purple stripes; the wings were orange and the keel dark purple towards the tip. The stamens were diadelphous, broader at the tip, with smaller anthers, and the ovary was quadrangular.

The northernmost cape of Öland has a strange shore, upon which the sea throws up large boulders annually, in the shape of ruins of fortifications and walls, and thus the cape elongates each year. Småland could be seen in the west from here.

Our way went from Öland's northern cape towards the south, for we followed the western shore as we had already followed the eastern.

A hare danced in front of us along the shore.

Woad (*Isatis s. Glastum*) [*Isatis* 543 *Glastum* of Index = *I. tinctoria* L.], a herb, which Professor Tillands formerly found near Abo and depicted, grew wild here upon the shore. The root is thin and survives only two years; it does not come into flower the first year; the root leaves are ovate, narrower at the base, with crenate edges, smooth and not rough. The stem is 1-2 ells [1-2 feet, 30-60 cm] tall. The stem leaves are like those of *Brassica perfoliatae vel Turritis altissima* [*B. campestris* L.], smooth and clasping the stalk. The inflorescence (*corymbus compositus*) has flowers with yellow, spreading petals. The fruits are pendulous. When this plant is chewed, it tastes of kale at first, then a sharp taste like cress. From this plant people make a blue dye, which is used for dyeing clothes blue. We showed the plant to the farmers who accompanied us.

The way went on the left hand into a wood, after we had followed the shore for some time and noticed now the *lantborg* on the western side could be recognized again; we had not seen it on the eastern side since we left Gaxa.

Serapias radicibus fibrosis, nectarii labio obtuso petalis breviore, foliis ensiformibus [*Cephalanthera longifolia* (L.) Fritsch], a plant which we have not seen before, and no one else in Sweden has either, occurred in the wild forest.

Radix perennis stolonibus decussatis. Caulis erectus pedalis simplicissimus: Foliis ensiformibus, strictis, striatis, quinquenerviis. Racemus erectus, terminatrix. Florum Calyx triphyllus, albus, erectus, corolla longior: Foliolis lanceolatis, carinatis, aequalibus. Petala tria: quorum duo ovata, conniventia, alba; infimum eadem longitudine, album, trifidum: Lacinia intermedia subcordata, obtusa, interius quinque striis elevatis exarata, extimo apice reflexa, villosa, lutea; Laciniae laterales erectae versus pistillum; basis pone gibba, nectarifera. Stamina duo in apice styli sub quibus Lacuna ad ovarium ducens.

The road to Torp through the forest showed us large numbers of the *Sanicula* [*S. europaea* L.] of the apothecaries and the *Pyrola* with a single flower. *Pyrola scapo unifloro* [*Moneses uniflora* (L.) Salisb.] was in full flower; we collected several flowers in order to be able to settle the argument among the Botanists about the number of stamens to each petal, but we found that although there are always ten reflexed stamens the numbers to each petal is so uncertain one looks in vain for a rule. The relations between stamens and petals were as between letters and numbers below:

A	B	C	D	E
1	2	3	2	2
1	2	2	3	2
1	2	2	2	3
1	2	3	1	3
1	3	2	2	2

We stopped in Torsby for a few hours at noon. The meadows were yellow with *Crista galli* [*Rhinanthus crista-galli* L.] and in the marshes grew *Ranunculus apium risus dictus* [*R.* 463 *risus sardous* of Index = *R. sceleratus* L.], *Pedicularis 1* [*P. palustris* L.], *Gramen aquaticum Fluitans* [*Glyceria fluitans* (L.) R.Br.].

Mercurialis caule simplicissimo [*M.* 823 *perennis* of Index = *M. perennis* L.] occurred here in every bush; it is harmful to sheep as well as to man.

Wild apple-trees grow all over Böda parish in all meadows and bear many apples, but the farmers know of no use for them except at the autumn slaughter, when a few apples are added to the meat soup. From these one could, as peasants do in England and Normandy, press out that beautiful cider or apple juice, which often competes with wine in flavour and is much more pleasant than the seawater, with which one saw the workers in the quarries quench their thirst.

The service trees [*Sorbus domestica* L.] were in full flower; the sorb-apples are eaten in the autumn, after the first frost; if they are eaten earlier, they are roasted first. The wood is not used for anything except for cogs in flour-mill wheels.

Primula 2 [*P. farinosa* L.] we noticed to grow only on acid soil, where the grass was short.

Flax is said not to grow here because the land is too low, the farmer had to say something when his womenfolk were reproached for their poor garments; very few have linen clothes here, and those who have them do not wear them.

Neptune's field is our name for a place ⅛ of a mile from Torp, on the seashore, one gunshot wide [225 m], a few gunshots long. This looked like the fields in Skåne or Uppland which have high ridges and deep furrows for better drainage; we could have solemnly sworn that nothing but a plough could have made such furrows, if we had not touched it with our hands and noticed that it was gravel from crushed spar, thrown far up on the shore by the storm and the waves of the sea, so that after the ebb of the water it was furrowed like a field.

The lime-kilns which here could be seen far from the shore, were usually 5 ells [10 feet, 3 m] high and 4 ells [8 feet, 2.4 m] broad; the two small furnaces were sloping inwards to facilitate the filling with fuel. We noticed that the upper stones were covered with something that looked like white hoarfrost; the people told us that it was ashes, but when we removed the hot stones we noticed that it was a white *sal alkali calcareum & fixum* which had sublimated like *flores sulphuris.*

The quarries from which this limestone came were situated on the shore, often a quarter-ell [15 cm] under water, where the limestone was quarried with a hammer and a chisel without much labour. When the blocks were separated there were as many "darter" or "Öland-spikes" [petrifactions] as there are husks in a coarse bread, and God alone knows where else so many rare shells could be found.

One petrifaction [*Entomolithus paradoxus*] *, which is rare in other places but common here, was similar to a *valvulam Echini,* often as big as the palm of a hand; it looked like a half moon with two parallel furrows and several transverse lines.

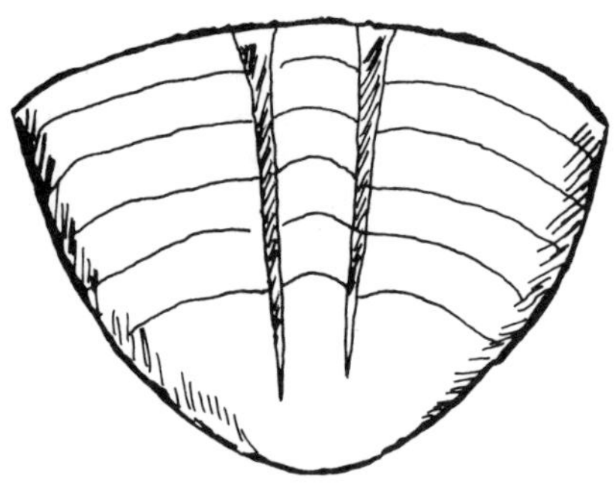

The spar crystals in the "darter" were similar to saltpetre crystals but had no *columna.*

The women, who drove oxcarts loaded with stone, were knitting stockings meanwhile, although they would keep a grip on the reins when they were actually driving.

Coccionella coleoptris rubris lineis quator albis longitudinalis [*Halyza oblongo-guttata* (L.)], an insect which is one of the biggest of its genus, jumped on the ground; both above and below it was pale red with long lines and white points on the wing-cases.

The buckthorn [*Rhamnus catharticus* L.] that grew close to the *lantborg* at a distance from the lime-kilns was the biggest one we ever saw; it was more than 3 fathoms [18 feet, 5.4 m] tall and the girth was 5 quarter-ells [75 cm].

The junipers had started to give off powder [pollen] and *Sedum acre* showed its yellow flowers.

The spruce forest gave us beautiful coolness as soon as we entered it, for sweat was pouring off us in the sunshine; all trees growing in a humid soil absorb large quantities of water, and this water cannot be excreted again through the roots but must evaporate from the pores and be dispersed into the air, thus cooling it. Doctors imitate this when they cool the air with water-plants when treating febrile diseases in summer.

The sandy moor we arrived at now produced only fir and tall heather, but no spruces could subsist on this dry soil.

When we left the road on the *lantborg* towards a little hamlet called Byrum we saw sandhills once more. Here was a whole district covered with sand for more than ¼ of a mile. The sandhills at Byrum are not as high as those in Grankulla which we saw in the morning. The sand was all white and smooth

---

* The petrifaction described by Linnaeus and named *Entomolithus paradoxus* by Schreber in 1764 (*Reisen durch Oeland,* 162) is the pygidium of a trilobite (*Megalaspis* species) cf. A. G. Nathorst, *Carl von Linné als Geolog,* 45, fig. 6. 1909.

and finer here than at the first place. The sand sedge [*Carex arenaria* L.] grew nicely on the sandhills as if it were planted along a string; at each section of the root new shoots come forth; this plant grows where the sand is densely packed and could be used for binding quicksand, since the roots are quite long.

Linnaea [*L. borealis* L.] was flowering in profusion between the sandhills and the village.

We could not get any food during the whole day: last year's harvest had been poor and the farmers did not have any bread for their own use; some had not had any for a month; others not for half a year, but after much persuasion we had a drink of milk, nothing else.

We lodged in Byrum, where we were plagued by mosquitoes the whole night, as if we had been in Lapland.

June 19 [p. 149]

Today was the third Intercession Day and we had to remain quiet and fasting; we spent the time on the sandhills which we left in the late evening yesterday.

"Sandpill" (*Formica-leo*) [ant-lion] is the name of an insect which few people in Sweden have seen; we amused ourselves with it. On the smooth surface of the sand-hills there were tracks, as if the thong of a whip had been loosely drawn over the sand, and at the end of the track there was a little pit, like the impression of the narrow end of an egg in the sand. When these pits were explored, we found the insect *Formica-leo* between the humid and the dry layers of sand. *Formica-leo* was as big as a spider, grey, with large oval belly upon which it had five rows of warts. The neck was slender, at the mouth were two claws, slightly curved at the tip, which it could open like a stag-beetle, these claws had three sharp teeth on the inside. There was hair on both sides of the claws and on the warts on the back. The feet were white. When *Formica-leo* was put down on the sand it walked backwards, digging itself down into the sand with its tail. The boys play with these ant-lions; they lie on the sand and, when they discover a pit, they blow forcefully in the centre of the pit to remove the sand and expose the ant-lion; those who fail to do this in one blow are ridiculed by their playmates. These ant-lions are the grubs of a fly (*Hemerobius*)* which lays its eggs in the sand, usually in the vicinity of an ant-hill with its ant-tracks. When the eggs are hatched, the grub burrows its way down into the sand until it has made a pit and then waits under its centre, protected from the birds. When the ants walk about performing their various tasks, they sometimes reach the brink of the pit and fall down helplessly, since their feet cannot get a hold in the loose sand, and they are attacked and devoured by the ant-lion. We put an ant into the pit, in order to observe all this exactly, and noticed that when the ant was struggling with all its force to get out, the ant-lion would throw sand at it, so that it slipped and fell into the lion's claws. When we put one ant-lion into the pit of another, the intruder was attacked at once. When we placed an ant-lion on the ant-track, the ants gathered to attack it, but the lion defended itself courageously with its claws as long as it had its back clear; it is more anxious to protect its tail than is the bear.

* The ant-lion is the larva of *Hemerobius formica-leo* L. = *Myrmeleon formica-leo* (L.).–*W.T.S.*

The shore, where these sand-hills ended and which was situated east of Blåkulla, had three characteristic plants:

Eryngium foliis radicalibus subrotundis plicatis spinosis, floribus pedunculatis [*E.* 220 *maritimum* of Index = *E. maritimum* L.] was one of those growing here on the shore; it has not been described in Sweden before. The roots go deep into the sand and the shoots which had not yet reached the surface were white and stout like asparagus and tasted good even when they were raw. The dried or preserved roots of *Eryngio* are sold in the pharmacies. The spears (*Turiones*) of *Eryngio,* cooked and eaten like ordinary asparagus, are a delicacy; they are diuretic, purify the blood and raise the spirits.

Salsola foliis pungentibus [*S. kali* L.] grew here and there; I have seen it before on the shores of Skåne.

Arenaria foliis ovatis acutis carnosis [*A.* 375 *maritima* of Index = *Honckenya peploides* (L.) Ehrh.] grew here with roots lasting from year to year. The stem-leaves were ovate, without stalks, smooth, succulent, with cartilaginous margins. The sepals were upright and hollowed. The five petals were outspread, ovate and shorter than the sepals. The stamens were ten and three short styles stood on the ovary, which was ovate, almost triangular, with one chamber; the whole plant smelled of fish.

The service-trees harboured a little leaf grub which penetrated into the inside of the leaf with its point; it was clad in a grey perpendicular cornet [cone], which on one side was smooth and black, but on the other side hairy and white.

The birch trees had uncountable numbers of grubs which devoured the leaves and made nests as big as hat crowns. We saw the same kind of grubs on the blackthorn at Torslunda on June 4.

Oaks, which were more numerous here than in the extreme north, looked as if they had been scorched; the leaves had been eaten by grey grubs with black spots.

We found a dead *Scarabaeus maxillis lunulatis prominentibus dentatis, thoracis utrinque tridentato* [*S. tridentatus* L.] on the shore. The head, the neck and the belly were dark brown with little excavated spots. The breast was hirsute below and prickly on the sides. The creature was as big as a stag-beetle but more rounded.

Papilio apiformis, abdomine aureo [*Aegeria apiformis* (L.)] was found here among the trees; it was like a bee in shape and size; the wings were white with black veins, the fore-wings were black in the middle as well as at the tips, but the hind-wings had only black margins. The antennae were likewise black, but stouter and white in the middle; the abdomen was black with two pale yellow stripes and a pale yellow belt and more hirsute and rust-coloured at the tip.

Along the sandbelt grew *Pulsatilla flore pendulo limbo reflexo* [*P. pratensis* (L.) Miller]. *Potentilla, quae quinquefolium majus repens C.B.* [*P. reptans* L.], *Pyrola floribus umbellatis* [*Chimaphila umbellata* (L.) Barton] grew in the wood and *Melampyrum spicis conicis laxis laceris* [*M. arvense* L.] in the fields.

The houses were built of rounded logs, for which the timber is cut in the seventh week before midsummer, when the fir sap rises and the bark is easily peeled off. The benches inside, 1 ell [2 feet, 60 cm] high, along the walls of the room, are hollow and filled with straw and covered with long cushions, to sit on during the day and recline on in the night; at the long table there is an equally long bench, like a sofa with a backpiece which can be turned to any side, so that the back can be supported when one is facing the table or the room. When the women are at their ease in the house, they wear a vest which is sewn to the skirt and can be hooked in the front when this is needed; they rarely use shifts under it and the naked body often shows under their garments.

Finally we were taken from Byrum by Superintendent Sahlsten, who on our behalf contracted with some farmers to ship us to Gotland by the first favourable wind.

June 20 [p. 153]

While we were resting and waiting for better weather in Horn, we spent the time looking for plants, insects and other less well known Swedish animals.

Curculio niger, elytris rubris, capite pone elongato [*Apodectus coryli* (L.)] ; this strange insect had red densely spotted wing-cases; the breast was red, smooth and narrow. The slender neck, the head, the tip of the breast and the belly were black. The head looked like a dog's head with a short snout; the eyes protruded; the outermost joint in the antennae was the stoutest, but not as long as on others of this species, the antennae were attached to the snout, and the claws on the feet were single and not double.

Cassida viridis ovalis laevis, clypeo caput tegente integro [*C. viridis* L.] found on the leaves of herbs. From above it looks like a single oval convex green crust, divided by a *sutura lambdoidea* into a cuirass and two wing-cases. The antennae are slender, somewhat club-shaped, dark but lighter at the base. When it lies on its back, one finds that the green casing encloses the black abdomen; once on its back it rights itself only with difficulty, spreading the wing-cases and holding on to the ground with its feet above its head.

A grass snake was killed; it had a grey back and a black belly, a row of white scales lay along each side of the body; under the chin and throat it was whitish, and on the white upper jaw at the ears light yellow, with transverse black lines. All teeth were of equal size, hence the snake is not dangerous.

Draba foliis caulinis numerosis incanis, siliculis obliquis [*Draba incana* L.] grew in the field; the root perishes annually; the leaves were ovate but those on the stem were heart-shaped, slightly hirsute with three teeth on each side. The flower spike was tall and the pods egg-shaped and compressed, with the septum between the chambers as broad as the outer wall.

In the afternoon we strolled through the meadows along the Royal Stables of Horn, where we found *Serapias flore albo* [*Cephalanthera longifolia* (L.) Fritsch] and a *Riccia.*

Riccia foliis aspergine crystallina perfusis margine incrassatis [*R. crystallina* L.], a very small plant, in all no broader in diameter than a pea; it was bright green and appeared to be sprinkled with little vesicles, almost like *Mesembryanthemum annuum*; the leaves and the branches were blunt.

Rana temporaria Charl. onom. 24 [*R. temporaria* L.] was caught and described; the hind feet had six slightly webbed toes, the first one small, the following of increasing length, the last but one being the longest. The fore-feet or hands had four toes, well divided, whereof the second and the fourth were the shortest. The back was flat and divided from the abdomen by an elevated line from the brows to the tail. It was all grey with black lines and warts on the back; the thighs were still paler with black transverse stripes.

A *Cochlea* found in the leafy grove was peculiar in that its shell was very thin, transparent and colourless; otherwise it was similar to the common dappled *Cochlea nemorum* [*Helix nemoralis* L.] ; when the white creature had withdrawn into its shell, this seemed to be decorated with branches and watery veins but when it crawled out of the shell, the veins disappeared; we also noticed that when it was in its shell, a little body fluttered and beat constantly, like the balance-wheel in a clock, which in all likelihood was the heart.

Asilus corpore atro glabro, alis nigris, femoribus claviculisque ferrugineis [*Asilus oelandicus* L.]. This predatory fly caught other small insects, tearing them into pieces with a big hard tooth under the jaw.

Cimex oblongus rubro nigroque variegatus, alis fuscis, maculis albis [*Lygaeus equestris* (L.)] was very like the insect that is common on the henbane [*Hyoscyamus niger* L.], with a red St. Andrews cross, only twice as big. The back was red, with the neck black and a transverse streak across the wings; there was a black rose between the upper branches of the cross but between the lower branches the wings were dark with white spots. The belly was red and on each side of it were five black points, and three other black dots longitudinally below the belly; feet, tail and antennae were black, the forehead red.

Alopecurus culmo erecto [*A. pratensis* L.] grew rather tall here; it is good for sowing in humid soil.

Other plants around Horn were especially *Epilobium foliis ovatis dentatis* [*E.* 306 *dentatum* of Index = *E. montanum* L.], *Epilobium foliis linearibus* [*E.* 307 *lineare* of Index = *E. palustre* L.], *Scutellaria 1* [*S. galericulata* L.], *Scrophularia 1* [*S. nodosa* L.], *Cucubalus floribus dioicis rubris* [*C.* 361 *dioicus* of Index = *Silene dioica* (L.) Clairv.], *Silene quae muscipula montana hirsuta* [*S.* 366 *muscipula* = *S. nutans* L.], *Lapsana* [*L. communis* L.], *Prunella* [*P. vulgaris* L.], *Stachys sylvatica, Verbascum nigrum.*

The farmers in this place, to my surprise, harvest only fourfold or fivefold the grain used for sowing, and yet the inhabitants of Falun can regain as much as eight- to twelve-fold. Perhaps the proverb (the better the soil, the worse the farmer) gives its cause?

Seals are caught in the parishes of Högby and Böda by means of seal-stones and upright or horizontal nets, made of horsehair. There were also other types of nets here.

June 21 [p. 156]

The gale today was as strong as yesterday, therefore we went to church in Högby.

At five o'clock in the evening, it calmed down; therefore we left Horn, passing Högby church, and after a mile's journey we arrived at the ship. On the road we saw a whole field of *Thlaspi 1* [*Th. arvense* L.] and much *Lithospermum officicarum* [*L. officinale* L.] at the lime-kilns.

There were crystal-apples in the burnt and unslaked lime; the fire had made them darker and they shone when broken like a spar.

We set sail at 9 in the evening, after waiting only a couple of hours for a fair wind. The southwesterly carried our ship from the land, which disappeared from our view at the same time as the sun, and our eyes were closed by sleep.

[On 22 June 1741 Linnaeus and his companions landed in Gotland, where they remained until the morning of 25 July, as related in Linnaeus's

Gothländska Resa. *From Visby, Gotland, they sailed on 25 July to Böda, Öland, stopping on the island until the morning of 27 July, when they sailed to Kalmar. Linnaeus accordingly inserted the account of this second visit to Öland, at the end of the account of his first visit, out of its chronological sequence.]*

July 25 [p. 157]

The sailors climbed up the mast, whence they could see Öland and Gotland at the same time; the main-yard was taken in, the foresail was enough. Öland came into view for all, the gale subsided, the ship anchored at Böda church and we praised God who had delivered us from the peril.

Then we proceeded to Horn farm where we spent the night.

July 26 [p. 157]

We were on horseback early in the morning.

The plants *Carduus acaulis* (July 19) [*Cirsium acaulon* Scop.], *Carduus nutans* (July 17), *Melica petalis exterioribus ciliatis* (July 19) [*Melica ciliata* L.], all three until now foreign plants, were common here.

We passed Föra church on our left hand.

A little further on we saw Pesnäs church once more (June 12).

Aster pratensis autumnalis conyzae folio Tournef. (July 11) [*Pulicaria dysenterica* (L.) Bernh.] grew abundantly in the furrows in the meadow close to Södvik Inn, which we visited on June 11/12.

Centaurium minus [*C. littorale* (D. Turner) Gilmour], short and bushy, grew here in spite of the long distance from the sea.

"Agh" [*Cladium mariscus* (L.) Pohl.] was said to grow in Källa parish west of Gatsjö village, in a bog there. The local name is åm.

The journey to Södvik was 1¼ miles and just as far as to Ormöga at Alböke. Alböke church was like all other Öland churches which are as plain as the Gotland churches are handsome.

Gryllus antennis longitudine corporis [*Stauroderus apricarius* (L.)], or grasshoppers, in millions jumped up and flew like dust before our feet everywhere over the *alvar*-land. They were of the same size as those seen earlier (July 10) but of another colour, paler than green. The upper side of head and breast were pale. The breast harness had a black spot on each side. The upper side of the abdomen was pale with black sides and yellowish brown underside. The thighs were pale with a long dark line.

Gryllus elytris nebulosis alis rubris extimo nigris [*Psophus stridulus* (L.)], a grasshopper soot-coloured all over, was seen among these, but was not so common; it is one of our biggest grasshoppers. The antennae are soot-coloured, thickest in the middle and half as long as the body. The fore-wings have a cloudy appearance, but the hind wings are nicely patterned with veins, like a net, and have wavy margins with tips black but red below. The thighs were black and pale yellow on the inside, the four front legs were dark and the back legs yellow.

Köpinge, which we had visited earlier (June 2/3), did not retain us, instead we proceeded to Räpplinge, 1½ miles from Alböke.

Räpplinge church and parish and its annex Högsrum are situated in the middle of the *alvar*-land and have lovely woods.

The fields were pale yellow now with the ripening crops, asking to be harvested any day.

The meadows had been mowed, and the harvest was smaller than last year.

The people told us that some cover the floors with *Galium luteum* [*G. verum* L.] on festival occasions, but it had the disadvantage that the guests started to quarrel and fight. This I cannot explain as a *Physicus* [natural scientist], but I will resolve the matter as a *Logicus* [logician], although the syllogism will be a quarter-ell long: *Galium* is never strewn on the floor except when there is a feast; here (as in other places) there is never a feast without drinking; there is never drinking without a fight; hence *Galium* promotes fighting.

Malva caule erecto foliis multipartitis [*M.* 582 *Alcea* of Index = *M. alcea* L.], or by common name *Alcea vulgaris,* grew in the meadows in Köpplinge and had not been seen in Sweden before. It is so common abroad that I need not describe it.

Vipetorpsborg between Köpplinge and Högsrum was visited. It is moderately large in comparison with the other ruins in this part of the country.

The woods were incredibly beautiful with lindens and the ground clothed with *Melampyrum coma coerulea* [*M. sylvaticum* L.], which by its yellow flowers and bright blue spikes lit up the shade.

There were several quarries along the road which were much deeper here than in other places, up to 1½ fathoms [9 feet, 2.7 m]. The people were said to neglect their other work here because of too much quarrying.

We had a quick look at Högsrum church, and from there we proceeded westwards towards Gärdslösa.

"Odens flisa" (Wotan's Rock) was the name of a pair of high flat upright rocks standing on a line along the road to Gårdslösa; both were about 1½ fathoms [9 feet, 2.7 m] tall and more than 1 fathom [6 feet, 1.8 m] broad, as close to each other as possible. This, supposed to be Wotan's tomb, was situated in a most delightful place; the land was high; in the west and the east were vast fields; to the north was a mountain ridge; to the south an oak forest, and around were several barrows and ancient tombs. Insatiable greed had not spared Wotan's grave; it had been plundered.

There were several ancient tombs, and not far north of Wotan's rock they were of a very peculiar construction, in the shape of a ship with the prow marked out with stones and the form of the ship outlined with 16 rows of smaller stones, like benches, and in the middle a couple of large stones. The ship was 32 paces long and more than a fathom [6 feet, 1.8 m] wide.

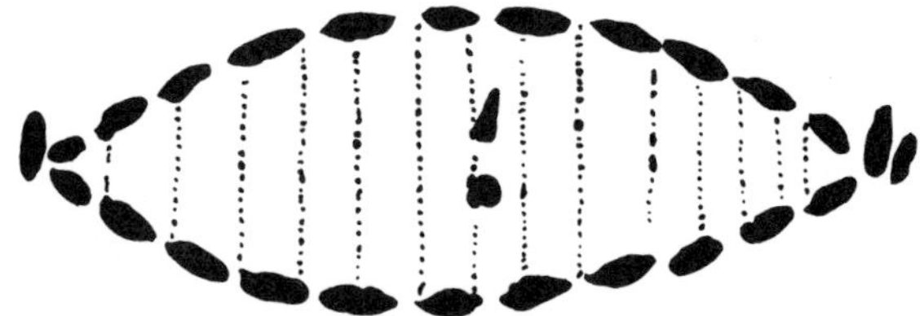

We returned to Isgärde from Wotan's Rock.

Stoechas citrina latifolia C.B. [*Gnaphalium* 674 *stoechas* of Index = *Helichrysum arenarium* (L.) Moench] abounded in the meadow at Brunderum; it has only been seen in Skåne before. It is hardly ½ ell [1 foot, 30 cm] tall, unbranched, with scattered hirsute and narrow (*linearia*) leaves. The inflorescence terminates the stem with its pale yellow shiny bracts and bright yellow flowers.

The "bog lime" or "allwar-moen" [*alvar*-soil], as it is called here, was usually brown, except at a few places in the valleys around Wotan's Rock where it was white. The reason for this is that the stone at Wotan's Rock is white, and the bog lime imitates the colour of its parent stone.

We were eager to see Algutsrum church and still more the ruins, which are said to be the largest in this part of the country; but evening and darkness came upon us and we had to hurry to Färjestaden (where we arrived on the first of June) where we stayed overnight.

July 27 [p. 160]

In the morning we crossed to Kalmar with the big ferry, in nice, calm weather.

There were thousands of jellyfish or medusae [*Medusa aurita* L.] in the water, and the water surface, which was calm like a mirror, looked like a sky with stars when the clear sunshine was reflected by the jellyfish. These were rounded, bulging upwards and hollowed from below and had fringed margins, in the middle of the underside were four horseshoe-shaped cavities, surrounded by an opaque curved line, which consists of little more than twenty parallel yellow grains; from the centre emerge four sickles (*falces*) with a ciliated outer margin; the whole creature is transparent like glass except for the four cavities, and its upper side is marked with veins radiating from the centre. This contracts and enlarges in the same way as a heart.

There were millions of mosquitoes floating on the water; they were of two kinds. When we tried to catch them they escaped; one was green and lifted its forefeet upwards, and moved them like horns (*Tipula pedibus anticis maximis antenni-formibus motitatricibius apice albis*) [*T. vibratoria* L.]. The other one was *Tipula thorace virescente alis membranacei coloris puncto nigro* [*Chironomus plumosus* (L.)] ; this was the one we saw in such large numbers at Färjestaden and in Ottenby Park.

Öland vanished out of sight, but its green meadows, its shadowy woods and incomparable *Tempe* stay clear in my memory.

Map of Gotland by Jacob Faggot
(reproduced from Linnaeus's *Öländska och Gothländska Resa*, 1745)

Biol. J. Linn. Soc., *5:* 109-220 plus Index. With 9 plates and 1 map

June 1973

The Journey to Gotland

CARL LINNAEUS

made in 1741

June 22 [p. 163] *

We awoke at dawn, two o'clock in the morning, and found the Karlsö islands before our eyes. The wind died down gradually, the ship gliding on; the sides of the Karlsö islands appeared more and more steep and high like fortification walls. The sea grew calm and smooth as a mirror, on which velvet scoters [*Melanitta fusca* (L.)] swam here and there; no porpoises were seen, nor any ships. The sailors passed the time with idle talk of two large carbuncles, said to have been taken from St Clement's church in Visby and gone to the bottom with the foundering of the ship of the Danish king Valdemar IV; they were the only ones to believe that the mast could still be seen when the sea was calm, and the glitter of the carbuncles caught by the sun sometimes shone into the eyes of seafarers; however, Klinteberg came nearer and Gotland's shore lay on our right hand, while the Karlsö islands were left behind, looking like limestone walls of considerable height. The sun shone warmly; time passed till 2 o'clock in the afternoon, when we landed at Visby.

June 23 [p. 163]

Visby is the only town in Gotland and is situated midway along the west coast. The town seemed to us like a model of Rome. So many, so large and such splendid churches stood all over the town, now roofless and turned into ruins by the changing times; their high walls [see Plate 4] were made of substantial hewn stone, without addition of bricks; their marvellous pillars and artistic vaults brought to our minds this town's former prosperity. The town occupies a semicircular space on the slope up from the sea. It was not very large, on the land side confined within a high wall, in which were many substantial and old towers; the wall is surrounded by two dilapidated embankments. The streets are mostly uneven and narrow with some irregular alleys; the houses are German, some of stone, some cross-timbered, and some of wood; most of them were roofed with tiles bought in Germany; some of the houses were so old that their walls were all black outside. Most of the houses,

* The page numbers following the dates refer to the page of the 1745 *Öländska och Gothländska Resa* on which the entry occurs.

which were inhabited and built of stone, have vaults on the lower floor, which were supposed to have been used for storage by the merchants.

The inhabitants were good-humoured, polite and friendly; their speech was somewhat different from ordinary Swedish and was somewhat like Norwegian in accent. The water, which comes from the land side, is clear, abundant and running, so one could have small pools with fish in the cellars. We did not notice that this water caused stone diseases [i.e. stone in bladder or kidneys] or podagra [gout], although it ran out of the limestone upon which the town is built, and although the tea-kettles became caked with a layer of chalk or *tophus calcareus* after long use. However, coughs were not uncommon among the inhabitants, and *Colica Hypochondriaca* plagued many; foreigners often get a rash on their hands when they first wash in this water, but the strangest thing was that the linen clothes of foreigners were said to turn red when washed for the first time in this water, then become whiter with every washing.

There were several uncommon plants on streets and in churchyards, as for example *Echium* [*E. vulgare* L.], *Cichorium* [*C. intybus* L.], *Conium* [*C. maculatum* L.], *Chenopodium 2:dum* [*C. urbicum* L.], *Scandix seminibus hispidis* [*Anthriscus caucalis* Bieb.], *Chaerophyllum caule maculato geniculis tumidis* [*Ch. temulentum* L.], *Lepidium foliis pinnatis integerrimis, petalis calyce minoribus* [*Hornungia petraea* (L.) Reichb.], *Hyoseris caule diviso nudo* [*Arnoseris minima* (L.) Schweigg. & Koerte]. *Scandix seminibus nitidis ovato-subulatis* [*Anthriscus cerefolium* (L.) Hoffm.] or the true chervil, which has never before been seen in Sweden, grew wild among the crops; but anybody collecting chervil for domestic use should beware of *Ethusa* [*Aethusa cynapium* L.] if he would keep his sanity, since both plants grow together and they are quite similar.

The giant bones exhibited in the cathedral as wonders are really whale bones.

The fish which hangs in the same church above the picture of St. George is a *Piscis malacopterygius, cauda bifurca, pinnis dorsi duabus, ani unica, cum altera intra hanc & caudum e regione posterioris pinnae dorsalis*: it was foretold that, when this fish had wasted away, doomsday would come; then this must not be far off.

A big and high stone-cross (Korsbedningen) [Valdemar's cross] stood outside the East Gate; its ring was carved with letters, though overgrown with moss [see Plate 5].

The day was hot and clear.

June 24 [p. 165]

On Midsummer's Day, when we went to church in Visby, one saw many graves strewn with all kinds of flowers by the friends of the deceased, which strewing is done every holiday, although in winter fir twigs are used instead.

Leafy huts were built by some people to-day, but most of the inhabitants had adorned their homes with sprays of oak foliage, for which we saw horse carriages with oak twigs brought here yesterday, since birch leaves are forbidden, and common men do not argue *a minori ad majus.* Boys and girls ran laughing the whole night to their playgrounds.

Gasterosteus dorso tribus spinis armato [stickleback, *Gasterosteus aculeatus* L.], or a little fish called "Spigg", was found at the seashore by the town.

Madreporae simplices, a kind of coral, were found in the sand some yards above the water-line in great numbers; some of them looked like small cones, others like small chalices.

Madreporae aggregatae, densely patterned on the outside with small stars and showing a net of *lamellis parallelis perpendicularibus et decussantibus* when broken, were not uncommon on these shores.

Conferva fusca ramosissima perennis grew commonly on the seaweed and on the stones in the water.

The stoves in this town were made of iron and were said to be imported from Norway.

The benches in the rooms were generally covered with one rectangular or many small square cushions, *ut molliter ossa cubent.*

June 25 [p. 166]

Although we had ordered horses yesterday in order to leave early in the morning, we did not get any until the afternoon; however we spent the time describing some sea-birds.

"Ard" [red-breasted merganser, *Mergus serrator* (L.)]. The bill is rounded with backwards-pointing sharp 'teeth'; the upper jaw was prolonged at the tip; head bluish black, with a long hanging crest. Throat whitish. Breast yellow (*testaceum*) in front. Abdomen white. Back greyish black. Outer pinions were black but inner white. Wings together were white with two black transverse stripes. Tail black. Tail-coverts with ash-coloured wavy markings. Feet and bill red.

Anas rostri extremo dilatato rotundatoque: ungue incurvo [shoveler duck, *Anas clypeata* L.]. This was same colour as a snipe (*Scolopacea*); wings grey with a shiny copper-coloured spot between two white transverse lines. The bill was rather broad and rounded at the end with a hooked tip and the teeth were like small scales.

"Mafwe" [mave] is the Gotlanders' name for *Larus* [common gull, *Larus canus* L.], which is as big as a hen, all white, except for wings and back, which are light grey. The first pinions are light grey below but black outside with a white tip; of the outer pinions, 1, 2 and 3 have a white spot on the black and a white tip; the inner pinions are whitish. Head, body and tail are white. The tail is entire and not cleft. The feet are yellow, webbed, with a short thumb [claw]; the legs are half-naked; the edges of the eyelids are bare and deep red, as are the corners of the mouth; the bill is yellow, convex almost as in a magpie. This bird was a male.

We took our course from Visby northwards along the west coast.

Bellis [daisies], which are used in pharmacies and which I have hitherto only seen on the plains of Skåne, grew here abundantly.

Plants: *Orchis bulbis indivisis, nectarii labio quinquefido,* etc. *It. Oeland. 45* [*O. militaris* L.]. *Orchis bulbis indivisis, nectarii labio quadrifido punctis scabro,* etc. *It. Oeland. 45* [*O. ustulata* L.] occurred in the meadows. *Lotus leguminibus solitariis membranaceo quadrangularibus,* etc. *It. Oeland. 143* [*Tetragonolobus maritimus* (L.) Roth] grew near the town, though sparsely.

Rubia cynanchica C.B. [*Asperula tinctoria* L. = *Galium triandrum* Hylander] grew everywhere in the woods as abundantly as *Aparine minima* [*Galium uliginosum* L.] on Öland. Gooseberry, hawthorn, dog-rose, blackthorn and *Asclepias* [*Vincetoxicum hirundinara* Medicus] were seen here and there.

The pastures were covered with bearberries, cowberries, *Cistus Helianthemum dictus* [*Helianthemum chamaecistus* Miller], junipers and pines, among which my flower [i.e. *Linnaea borealis* L.] was sometimes seen, but neither heather nor fir grew in this dry soil. Many fields were yellow with charlock, and the meadows were yellow with *Ranunculus acris.* The true chervil was growing on arable land. *Trifolium lupulinum* [*T. aureum* Poll.] on the fields, and *Cotoneaster* [*C. integerrimus* Medicus] and Jasione [*J. montana* L.] on the hills.

Korpeklint was situated ¾ of a mile from the town and was regarded as a mountain in this place. The land slopes almost vertically down to the sea, as in Öland's *lantborg* and often as steep as an Attestupa [ancient "suicidal precipice"] ; the rock is pure limestone, in horizontal strata, grey on the outside but all white when broken, almost transparent like quartz, without grains, like a white marble. Korpeklint was likewise of such a nature; here we also found many petrifactions. The journey, which from Visby to Korpeklint had proceeded beneath the precipice, now took us on to its verge.

The road, which was entirely white, irritated our eyes when the sun shone upon it, as limestone, when worn to pieces and disintegrated by weathering, yields a white earth. The road, which was full of chips, was built and repaired with small rounded rubble-stones [pebbles] that were to be found everywhere and used instead of gravel. These rounded stones proved, although they were lying so many feet up and along the ridge, that the waves of the sea had rounded and cast them here all together. One can imagine that, when the sea first laid the foundations of the land, this ridge had been cast up as a bank, to which the sea daily cast gravel and sand and afterwards withdrew. The east shore should then be broader, as the waterfront on this side is so much longer than on the west shore.

Överstekvarn in Lummelunda parish was situated 1¼ miles away from the town, along a peculiar stream that here forms a waterfall for a flour mill, a sawmill and stamping machinery, so high that it is one of the highest in Sweden and certainly the highest on Gotland. Formerly, a blast-furnace and a hammer are said to have been driven by this stream.

The stream of Överstekvarn is very peculiar in that, from its source in the lake of Martebo, it runs underground for $^1/_{16}$ of a mile under hills and valleys, and reappears at Överstekvarn, where the *lantborg* on the west side is broken away, and where it flows out of a small vault, 12 feet wide and 6 feet high.

One and three-quarter miles from the town [Visby] we passed Nygranne Inn; then Lummelunda church, annexed to Martebo; and not far from here we found a plant, growing abundantly, that has never before been discovered in Sweden.

"Rams" [ramsons, *Allium ursinum* L.] was the name of the plant which here grew under the bushes; the root was an oblong bulb, surrounded by bristles. The leaves were lanceolate and in shape similar to the Lily-of-the-Valley. The stem was 6 inches [15 cm] long, quite naked, on one side flat, on the other rounded; it ended in a bract (*Spatha*) which is white, within which many flowers arose from one point, each with its separate stalk. Petals 6, lanceolate, white and outspread. Stamens 6, half as long as the petals. Ovary consisted of three round kernels with a simple style. The farmers say that where this plant grows, it drives away other herbs and weeds: we had the proof of this before our very eyes, since under those bushes where the ramsons grew there were no other plants. The farmers also told us that they plant it among the hops, to keep wild chervil [*Anthriscus sylvestris* (L.) Hoffm.] and other weeds away. The cattle eat it willingly, but it gives a garlic taste to milk and butter. This bulb or rootstock is closely related to *radix Victorialis longae* and could easily replace it for pharmaceutical purposes, since *Victorialis* [*Allium victorialis* L.] is difficult to obtain from the Swiss Alps and is most often without taste and smell when it reaches us. Botanists call this plant *Allium foliis lanceolatis scapo nudo semicylindraceo, bulbo setis obvallato* [now *A. ursinum* L.].

"Agh" [sedge, *Cladium mariscus* (L.) Pohl] is the farmers' name for a tall grass that grew in the marshes near Lummelunda church so abundantly, like crops in a field, to a height of 1½ ells [3 feet, 90 cm]. This was a grass that no botanist has found in Sweden before, and it is called in Latin *Schoenus culmo tereti, foliis margine dorsoque aculeatis.* The stalk was rounded or slightly triangular. The leaves were like those of the reed [*Phragmites*], very pointed, quite straight, palm-like and longer than the stalk. On the inner side smooth but on the margins and keel, saw-like with fine teeth. People thatch with this grass most of the roofs at their farmyard; it is mown with a scythe between midsummer and the feast of St. Olaf and is bundled while it is still fresh, without regard to which end is up or down; roofs thatched with this are better and tighter than those covered with straw. Those who do not have this sedge on their own land buy it where it grows and harvest it with their own men for a price of 8 à 16 "öre" a cartload. All of the marsh where the sedge grew, and which was several gunshots in diameter, had been a tarn or small lake formerly, although the water did not cover our shoes now and it was difficult to put down a pole in it: the creeping roots of the sedge had covered it so densely. How many sterile and useless bogs are there in Sweden that cannot be drained or made fertile in any other way? I think it would be worth while to plant sedge in them, and I have no doubt that the sedge would manage well once planted, especially since I have noted in the Academy garden that it withstands our winters. The cattle eat it only during the early spring, when it first comes forth, but the dry sedge is eaten only when there is an extreme shortage of fodder, and then it must first be threshed and winnowed, so that the small sharp teeth on the leaves do not injure the mouths of the cattle.

Orchis morio mas, foliis non maculatis. Bauh. pin. 81 [*O. mascula* L.] was growing in the meadow not far from the sedge and has not been discovered before in Sweden. *Radices testiculatae, ovato-oblongae; Caulis teretiusculus, erectus, strictissimus, infra flores pedalis. Folia 4 vel 5, liliacea, superiore latere glaberrima nitida, ultimo folio spathaeformi. Racemus terminatrix, spithameus, floribus circiter 20 instructus; singulis floribus: Corolla incarnata, versus faucem pallida; petalis obtusis exterioribus reflexis; 2 vero interioribus conniventibus in galeam. Labium inferius versus basin punctatum, extrorsum dilatatum, inaequaliter crenatum, trifidum: intermedia minore; lateralibus retroflexis. Nectarium cornu ascendens, germine fere brevius, obtusum, non vero emarginatum.* It is evident from the description, and especially from the shape of the flower, that this plant must be a variety of *Orchis bulbis indivisis, nectarii labio quadrifido crenulato, cornu obtuso* [*O. morio* L.], although at first sight they appear very dissimilar in general appearance as well as in the colour of the flowers. The farmers here call all orchids "S. Johannis-Nycklar" [St. John's Keys].

Gates here are built in a special way. They consist of two posts, put down into the earth, with horizontal rails inserted loosely into one post and fastened into slots in the other by a slat hung on two hooks, so that the rails cannot be removed until the slat is lifted away.

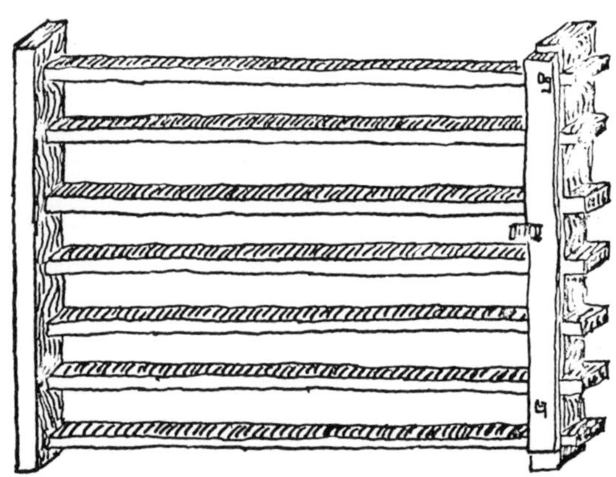

We had Martebo Träsk on our right hand before we arrived at Martebo church; this little lake was rather long, not so broad, but not at all deep, so that one could have waded across it at some places with safety, if there had not been patches of mud here and there on the otherwise very beautiful sand bottom. On the other side of the lake we saw some green tussocks, and we wanted to know what kind of grass was growing so handsomely there; thus we borrowed a couple of punts, which were well suited to the water. The punts looked like boxes with a flat bottom and perpendicular sides, 5 ells [10 feet, 3 m] long, 1½ [3 feet, 90 cm] broad, 18 inches [45 cm] deep, oval. Neither oars nor rudder were used, only a stout pole, to go to sea on this lake. Two persons filled up the boat, and they returned with the "Lapp's shoe grass", a *species Caricis* [*Carex vesicaria* L.] described in *Flora Lapponica*, n. 328. On the shore in the wet sand we found some *Staphylini.*

Staphylinus niger, elytris antice griseis, pedibus rufis [*Staphylinus littoreus* L.; cf. p. 13]. It was no bigger than a louse, but thinner, all black except for the wings, which were glaucous and very short. The tail was not cleft and was often lifted up.

Staphylinus rufus, elytris coeruleis, capite abdominsque apice nigris [*Paederus riparius* (L.)]. It was big as an ant, flesh-coloured; but the head, the short wings, the tip of the abdomen and the outer antennae were black. The jaws were pointed with a tooth on the inner side. The antennae consisted of IX oval jointed sections with the smaller end towards the head, except for the outermost section, which was smaller outwards.

Marrubium [*M. vulgare* L.] was growing near the villages, and we saw that twigs had been taken from the yew trees in this district: almost all the oaks had no leaves, since they had been eaten by grubs.

Collar harnesses were used for the horses here, when they drew carriages; they consisted of a transverse pole with a ring attached to its middle; the centre pole was put through the ring, the transverse pole lay across the breasts of the horses, in front of the shoulder, and was suspended by two straps around the withers of the horses, who must thus have strong necks and be well trained in order to draw with such a harness.

The roofs of the farm houses here are mostly made of boards, laid vertically from the ridge of the roof to the eaves without any birch bark or turf beneath.

"Riksdaler" is the currency most often used by the people and other currencies are counted as ½ and ¼ riksdaler, but this riksdaler is not worth more than 4½ "daler Kopp:mt" or a double *Carolin*; the reason for this is that it is counted in the same way as Danish crowns.

Martebo church, ⅜ of a mile from Nygranne, we reached at 8 p.m. and rested there after a hot day.

June 26 [p. 174]

In the morning we went out to botanize in Martebo. In the fields and on the roads grew *Anagallis flore rubro* [*A. arvensis* L.], never before seen in Sweden except in the vicinity of Lund on the plains of Skåne. The fields were yellow with charlock and we often saw the ground covered with a blanket of yellow *Ranunculus seminibus aculeatis. It. Oeland.* 104 [*R. arvensis* L.]. *Ballota* [*B. nigra* L.], which has not been seen before in Sweden either, except in Skåne, grew by the fences. *Ophioglossum* [*O. vulgatum* L.], from which a healing ointment was made, was called "läketunga" [healing tongue]. *Geum floribus nutantibus* [*G. rivale* L.] was called here "katballar" [catballs] and *Cyanus* [*Centaurea cyanus* L.] "båtzmans-hatt" [sailor's hat] from the colour and structure of the flowers.

Household remedies were:

Valerianae radix [root of *Valeriana officinalis* L.] was used for hysteria; for the "rose" [erysipelas] was used sulphur with pepper; and for the ague, gunpowder, strong liquor, snuff and tobacco as an emetic, sometimes with addition of lukewarm water. *Semina Cardui Mariae* [fruits of *Silybum marianum* (L.) Gaertner] are here called "styngkorn" and ingested for stitch.

Yellow colouring is here got from "gietbarck" [goat bark] (*Rhamnus catharticus*) and "gulbark" [yellow bark] (*Frangula*) [*Rhamnus frangula* L., *Frangula alnus* Miller], of which the dried bark is boiled after maceration in weak lye, or a mixture of lye and water.

Maypoles were erected by the farm girls in the church as a midsummer decoration; they were beautifully covered with flowers of several kinds and chips of horn, snow white as fresh parchment.

The journey proceeded from Martebo to Stenkyrka.

The fields were clayey; wheat, rye and barley were sown, but neither flax, hemp nor peas, and we noticed everywhere that the people were not used to handling flax. Barley is sown for two years, the third year the field lies fallow, then it is manured, and rye is sown the following year, then barley without additional manuring. Seaweed (*Fucus*) was also, though rarely, used for manure in spite of the soil being clayey. The fields were manured with the seaweed, but it did not last as long as dung.

Charcoal stacks were seen burning in the forest; the tar wood was nothing but old pine stubs dug out of the earth, which stubs are supposed to be the remains of old, mature pine trees, which do not rot, but absorb more resin the longer they are left in the earth, so that a stub that has been left in the earth less than 30 years after the tree has been cut down is not considered to have taken up enough resin. I leave this paradoxical hypothesis for further investigation. I cannot understand, according to modern *principia vegetationis*, how a pine tree root that never puts forth any shoots could absorb any juice or resin when the trunk is cut off and it is dead.

At some farms we saw stone houses with tiled roofs; the walls were made of limestone with a mortar of mixed sand and bog lime, but the upper storey or at least the gables were cross-timbered. The farmers in this district never heated their houses with ovens, neither did they have dampers in the flues—thus one must have a steady fire during winter.

There were no hills here and the common bed-rock was granite.

The sedge marshes harvested last summer were still bare, since it takes several years for the sedge to regrow to full size, once it is cut.

Here as in most places the fences are made of sloping wattles and upright posts, but each pair of posts is joined with osiers at the top, two for each pair, never more, so that each wattle is supported by osiers at one end while the other rests on the ground. The top twigs are left on the osiers and the twigs are arranged so that they point in opposite directions. This makes the fence look as if it was covered with bushes, to prevent the animals jumping over it.

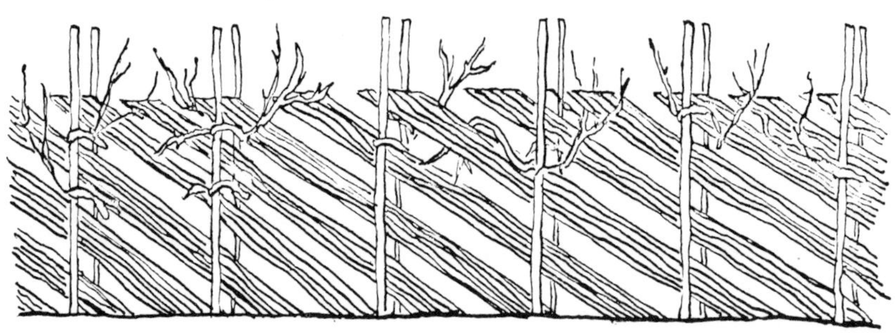

Stenkyrka did not show us anything very remarkable, but after a further journey of ¼ mile we found on the left hand side a limestone rock, where we discovered some plants which have hardly ever been seen in Sweden.

Cistus caule procumbente, foliis alternis [*Cistus* 435 *fumana* of Index = *Cistus fumana* L. = *Fumana procumbens* (Dunal) Gren. & Godron] grew here on the limestone rock; I have only seen it before at Fontainebleau in France. It is easily distinguished from other kinds of *Cistus,* since the leaves are not paired but scattered. *Caulis suffruticosus, diffusus, terrae adpressus. Folia alterna linearia, plana, subtus convexa, punctis minutissimis & vix conspicuis adspersa, marginibus minime reflexis; Calyces pentaphylli: foliolis exterioribus duobus viridibus, reliquis tribus striatis. Pedunculi uniflori; Antherae luteae; corolla lutea nondum explicata erat.* We did not find it anywhere else.

Anthericum foliis planis, corollis planis deciduis [*A.* 267 *ramosum* of Index = *A. ramosum* L.]. *Radix horizontalis, fibrosa, fibris carnosis; radicula centrali praemorsa; apex radicis barbatus, adeoque perennis. Folia radicalia numerosa, pedalia, graminea, linearia, acuminata, glabra, subtus levissime carinata. Caulis pedalis, nudus. Flores nondum explicati.* This grew fairly generally in the loose limestone; later on, we found it at Torburgen and other places.

Sedum foliis subulatis confertis adnatis, basi membranacea soluta, umbella racemosa [*S.* 388 *rupestre* of Index = *S. rupestre* L.] . This plant had not been discovered in Sweden before, but later it has been seen by the studious Dr. Leche [Johan Leche, 1704-1764] in Skåne and the industrious Herr Kalm [Pehr Kalm, 1716-1779] in Västergötland. *Caules simplices, erecti spithamei, racemis reflexis nondum explicatis. Folia alterna, subulata, carnosa, sessilia, basi subtus solata. Stolones plurimi, capitati, cespites ad radicem constituentes.*

Ruta muraria [*Asplenium* 855 *ruta muraria* of Index = *A. ruta-muraria* L.] , a *species Asplenii,* used in pharmacies and hitherto imported from abroad, grew on these bald rocks, *cum lineolis pulverulentis dorsalibus.*

Poa panicula secunda coarctata culmo obliquo compresso [*P. compressa* L.*] was very common here in the gravel; each spikelet consisted of 7 small flowers.

"Lappskogräset" [*Carex vesicaria* L.] with its long beautiful leaves was seen again growing in puddles; it should be planted in marshes, for no other grass grows more luxuriously; below the male spikes, which terminate the stem, one very often found some female flowers.

Immediately beyond Stenkyrka the limestone began again, as could be deduced from the white stones on the road; when limestone weathers, it turns white, but granite does not. The whole stretch was of limestone rock, which was no higher than the soil, entirely white, smooth and bare, with no covering of mould.

"Bleke" was found among these limestone rocks, and where the soil level between the rocks was somewhat lower one saw little dried-up marshy areas with white earth which is commonly called "bleke". Here one could see how "bleke" is formed in nature. These limestone rocks consist of a limestone, in colour like sugar with transparent grains, which is weathered by air, heat, cold and water, so that its surface becomes porous, grey and white; the loosened particles are washed off by the rain and flow with it into the puddles; when the water evaporates, the solid particles are left in the puddle and produce the "bleke" [bog lime] . Small pieces of limestone lay in abundance in the midst of the "bleke" and made us wonder whether the bog lime could also be reduced to limestone, but since we noticed that this gravel was only deposited at the margins of the puddles, we understood that the spring flood from the melting snow had carried these small stones with it; while the fine impalpable dust, mixed with the water, was carried further, the stones remained in the puddles when the stream lost its force, for which reason the bog lime is purer at the bottom of the puddle. Thus "bleke" [bog lime] is really lime and not chalk.

The road, which lay on this smooth, bare limestone, put us in very great danger of fractures as we rode over it, since the horses stumbled on it as on slippery ice without being able to get a foothold; for limestone has the property that the more it is worn down the smoother it becomes. It is also futile to put sand on it, since any sand brought here in summer would be washed away in spring and autumn.

* To the list on pp. 9-12 of species having Öland or Gotland as their restricted type-locality should be added *Poa compressa* L., *Sp. Pl.* 1: 69 (1753), *Fl. Suec.* 2nd ed. 28 no. 84 (1755), with Gotland as its restricted type-locality (cf. *K. svenska VetenskAkad. Handl.* 1741: 184; *Öländ. Resa* 178: 1745).—*W.T.S.*

The firs had at the tips of their twigs small cones like strawberries which consisted of needles with an enlarged base, lying like scales on top of each other, between which were a lot of small insects, which could hardly be seen with the naked eye. They were very similar to a kind of insect that is called *Aphides alni* and lives on alder twigs before its metamorphosis, but those on the fir trees were 100 times smaller; from their anus issued a kind of wool or stuff like transparent swollen gut; this insect is called *Kermes abietis* [*Chermes abietis* L. = *Adelges abietis* (L.)].

Bromus culmo indiviso, spicis alternis subsessilibus teretibus [*B. pinnatus* L. = *Brachypodium pinnatum* (L.) Beauvois] or "Sparrlosta" was also growing here. *Spica composita e spiculis distice alternatim enatis sessilibus*; the bracts of the spikelets were not towards the stalk as in *Lolium* but on the sides. Each *spicula* was *teres, disticha calyce diphyllo; squamis lanceolatis acutis; altera harum in aristam brevem desinente; flosculi 5. vel 6. alternatim distice positi; Flosculi singuli constabant ex duabus squamis, quarum exterior quinque nervis erat instructa, aristata arista terminatrice.*

Thalictrum caule folioso aequali, foliolis caulinis acutis panicula simplici, floribus nutantibus [*Th.* 254 *arvensis* of Index = *Th. minus* L. var. *kochii* (Fries) Neum.], a plant which we had seen hitherto only at Växjö, was growing wild near the river Ihre. *Florum panicula ramulis nutans, corollis tetrapetalis, fuscis; Antheris & filamentis corolla longioribus, flavis, Germinibus sessilibus; Caulis teretiusculus striis obscuris elevatis saepius septem, fusco purpurascens. Folia supra decomposita, foliolis tricuspidatis obtusiusculis.* The river Ihre passed under the road, and we found several natural history specimens in it.

Tremella tuberculiformis solida rugosa [*T.* 1021 *saxea* of Index = *T. verrucosa* L.] or stone plants were found on the stones under the water in the stream; when removed from the stone, it was rough at the base, but outside was smooth, slippery, dark and the size of a pea.

I call *Coccus aquaticus* [egg of *Hirudo octoculata* L. = *Nephelis octoculata* (L.)] a kind of oval raised body looking as if it were made of parchment, no bigger than a dill seed, which was also on the stones under the water, and resembled *Coccus hesperidum* [*Lecanium hesperidum* (L.)], the lice found on the stalks and leaves of lemons in orangeries. This little shell, which was taken from the stones, was covered below by a thin transparent membrane, so that one could see small elongated worms crawling within the body. At one end of this oval body stood a dark raised point but at the other was a gelatinous white two-forked beard. Here one found empty shells of this kind, with holes on the back that the worms had bored and escaped through. I leave it to others to investigate whether this is a *species Cocci* or merely the ovarium of some water worm [cf. Linnaeus, *Syst. Nat.* ed. 10, 1649 (1758), T. Bergman having demonstrated in 1756 that this was the egg of *Hirudo octoculata*].

Besides these the following insects were seen in the water; *Podura aquatica nigra* [*P. aquatica* L.] was jumping on the water together with *Cimex linearis supra niger pedibus anticis brevissimis* [*C. lacustris* L.]; *Dytisci* and *Nepae abdominis margine integro* [*Nepa cinerea* L.] were moving under the surface.

Hirudo depressa alba lateribus acutis [*H. octoculata* L. = *Nephelis octoculata* (L.)] or a white oval leech, in shape like the milt of a fish, was found under the stones in the water; it is often found very large in the stomachs of small fishes, and I would believe that the worms found in the livers of sheep which the peasants call "ilar" are nothing but the spawn of this leech, since the sheep pick up these liver worms while grazing in marshy places.

Phryganae were lying on the bottom of the stream, rolling in their cylindrical tubes made out of sand.

Gasterosteus aculeis in dorso decem [*G. pungitius* L. = *Pungitius pungitius* (L.), but possibly misidentified] was jumping in the water.

Papilio hexapus: alis erectis, rotundatis, albis; venis nigris [*P. crataegi* L. = *Aporia crataegi* (L.), the black-veined White] was also found here; it had on the fore wing a black spot at the *Anastomoses vasorum* which one would not notice if one did not examine it carefully [see p. 13].

We have seen "mjölonris" [bearberry, *Arctostaphylos uva-ursi* (L.) Sprengel] more plentifully in every bare patch in the forest today than in any other place in Sweden, so that should anyone desire it in great quantities, no-one could get it more easily than in the districts we passed through today and yesterday.

This hot day ended with the sun setting in a wall of clouds. After a journey of 1¾ miles we had to lodge for the night near Hangvar church.

June 27 [p. 182]

At 4 o'clock in the morning we rose to read four runic stones lying close to each other in Hangvar churchyard.

Lichen crustaceus candidus crassus, tuberculis atris [*Lecanora calcarea* (L.) Sommerf.] or a white lichen, which hindered so much of our reading of runic stones on Öland, was noted to grow predominantly on limestone, which made it possible to distinguish limestone from granite at a great distance.

A very big ash tree was growing by the church wall at Hangvar, of which the trunk had been torn on three sides from the root upwards as far up as we could see; the rents in the bark were healed with raised scars of a finger's thickness. Old farmers still remembered the flash of lightning that had caused the scars.

The baking ovens at the farms were always built outside the houses and covered with a shelter; the oven door and the fireplace were inside the house; this was to save space within the houses.

Alder is so uncommon in this district that we have only seen it at a few places and never many trees.

The fields were full of limestone shingles, which were said to cool the soil in strong heat.

Charlock was growing abundantly among the barley, but strangely enough not among the rye; it is difficult to figure out the reason for this; although the charlock drops most of its seeds before the barley is harvested, it grows neither the next year when the field lies fallow nor the following year, when rye is grown, but appears again in the third year, when barley is grown once more.

The grain is winnowed here after threshing; this is done when there is a strong wind; then the doors of the granary are opened, and a man standing in the doorway facing the wind shovels the grain; this makes the kernels fall down at his feet, while the lighter chaff is blown further away on the granary floor.

Hall is attached to the parish of Hangvar, situated ¾ of a mile westwards on a promontory. A chapel belongs to this parish, between Hall and Kappelshamn, in which chapel services are held in spring and autumn for the fishing village nearby.

Seals* are caught here on the coast mainly in two ways: with "lying nets" or "standing nets". Seal stones are big stones lying close to the water-line, a little above the surface, smooth and large; where there are no such stones, they are put there by the farmers. The standing nets are placed in a half circle around the seal stone, for if the seal is going to rest or sleep, it has to find such a stone, upon which it always climbs from the landward side and rests upon it with its nose towards the sea; when the seal is frightened, it quickly makes for the sea and is caught in the net. The lying net is placed like a four-cornered fence around the seal stone; its outer and inner ends are held open by transverse rods. One end of a long rope is tied to both ends of the innermost rod, the other end is left on the shore; when this rope is pulled, the rods, and the net which is attached to them, are drawn to the bottom, but when it is released the upper end of the net lies just at water level. The lower part of the net is attached to the base of the stone. When the net is down, and the seal climbs the stone in order to rest, it does not notice the trap; when the farmer sees that the seal is asleep, he pulls the rope slowly, which makes the net rise and surround the stone. When the seal wakes up, it jumps into the water and is caught in the net; it cannot escape before the man kills it with a harpoon or some other instrument [see Plate 6].

Flounders are caught in some quantity by fishing tackle or trotline (långrev), which is also called "ålrev", for which herring cut into pieces is used for bait. We noticed that these flounders were of two kinds: the larger [*Pleuronectes maximus* L. = *Scophthalmus maximus* (L.), turbot] has eyes on the left side, which is rough, grey and without red spots, of which the rays were in the dorsal fin 59, in the pectoral fin 12, in the abdominal fin 6, in the rump fin 39 and in the tail fin 16. The smaller variety [*Pleuronectes flesus* L. = *Platichthys flesus* (L.), flounder] has eyes on the right side, which is grey with flame-coloured spots, with rays in the dorsal fin 57, in the pectoral fin 9, in the abdominal fin 6, in the rump fin 39 and in the tail fin 16.

Sedge [*Cladium mariscus* (L.) Pohl] was used as bed straw, because all other straw had been given to the cattle as fodder during the winter; the maid who makes such a bed should not have delicate fingers, if she does not wish them injured.

We saw several limestone quarries along the road from Hangvar which were not deeper than 2 ells [4 feet, 120 cm] and were similar to the "alvar"†-quarries on Öland. The limestone was almost horizontally stratified and light grey, but somewhat rusty red in the fractures or cracks. In this stone are found numerous petrifactions, especially *Entrochi* and a kind of coral that looked like *Lycopodium.*

The plants that occupied this district were *Rubia cynanchica* [*Galium triandrum* Hylander], *Geranium pedunculis unifloris* [*G.* 571 *uniflorum* of Index = *G. sanguineum* L.], bearberry [*Arctostaphylos uva-ursi* (L.) Sprengel], *Cistus helianthemum dictus* [*Helianthemum chamaecistus* Miller], *Sedum teretifolium album C.B.* [*S. album* L.], *Sedum parvum acre I.B.* [*S. acre* L.] and sedge [*Cladium mariscus* (L.) Pohl] in the marshes.

---

* Common seal, *Phoca vitulina* L. In Syst. Nat. 10th ed. 1: 38 (1758) Linnaeus mentioned its habit of sleeping on rocks, having presumably learned this from his Gotland journey: *Dormiunt in lapide ex aqua eminente.* – *W. T. S.*

† Alvar is the Swedish name for a steppe-like limestone plateau in Öland; see pp. 3, 48.

Kappelshamn is situated at a distance of ¾ of a mile from Hangvar church where a large bay cuts into the land from the north-west and makes a good harbour for seafarers in all weathers except in a north-westerly wind, which often brings disaster. In Kappelshamn is a lime works, which has the lime kiln on the western shore by the seaside. The kiln is built almost like a blast furnace, 4 fathoms [24 feet] high, 2½ fathoms [15 feet] broad. The stones of which the kiln is built are renovated on the inside only. At "A" the unburnt limestone is introduced into the kiln and in the small fireplace at "B" fuel is put in and a fire lit; the opening at "A" is closed during the burning, which takes two days, during which time 50 to 70 lasts* are burnt. The burnt lime is then transferred to a special building on the shore where it is slaked and stirred. During the lime-burning the air is fuller of lime particles than any mill is full of flour dust; this sticks to the lips of the workers, causing rashes and blistering. The area surrounding the kiln was littered with waste limestone and firewood. The manners and the clothes of the owners reminded us of foundry proprietors. One owner of a lime oven can sell 700-1000 lasts every year. The whole place was full with wood for lime-burning. There is enough fuel here, but the wood is uneven in quality; if the lime-burners are allowed to do as they please with the woods, this part of Gotland will soon be cleared of forest, since for every 2 days' burning, 20 fathoms of firewood are used.

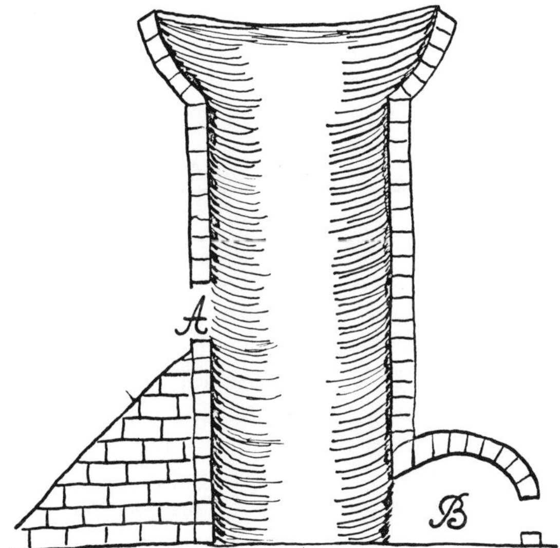

Stacks of beams were lying in quantity on the shore; they are sold mostly to Germans and Danes. Some beams were 14-16 ells [28-32 feet] long, 7 inches broad and 7½ inches high and cost 16 "styver" each; others were not more than 12 ells [24 feet] long, 6 inches broad and 5½ inches high and were sold at a price of 8 "styver"; most of the beams were split, because the farmers saw them to the right size immediately after the spring felling.

We collected petrifactions on the western shore for some hours; there were a

* Last: commercial measure which seems to vary with the commodity: a last of wool = 12 sacks = 4368 lb; a last of malt = 10 quarters = 80 bushels.—*M.Å.*

good many *Conchitae striatae* with *Cochlitae* among them, some simple, some gilded with a hood of pyrites or filled with crystals.

A sea-mist could be seen like dense smoke at the other side of the bay while we were refreshing ourselves at 12 o'clock with the customs officer Kiörsner. This mist came in from the sea and was borne past us into the bay by the wind, and did not lift, although the day was hot and it was noon. Finally the mist became so thick that it was like snow in the air and hid the sun from us by forming thin flying clouds. We were only a few gunshots [one gunshot = about 225 m] distance from the bay and whereas we saw the sun shining, the sea and the bay were covered by a heavy dark-blue mist, like the smoke that is to be seen over the copper mines at Falun; we saw a similar mist rising from the sea on the east side of Gotland. A few hours later, these mists covered all the visible part of the sky. From this we judged the great quantity of cloud and consequently rain that is evaporated from the sea for the inundation of the earth. The farmers told us that in their opinion this was the winter cold rising from the sea, where it had been hiding till now.

We left Kappelshamn and continued at 2 o'clock to Fläringe, where some of us crossed the bay in boats, while the rest rode around it with the horses; all parties arrived at the same time on the opposite shore, in 20 minutes.

"Corall-stranderne" [coral shore] I call that which lies to the eastern side of Kappelshamn, which was very broad and covered with white and grey stones; this greatly surprised us, since each one stone was nothing but a coral, called *Madrepora,* so that anyone who wants exquisite corals for collections of natural history specimens need try no other place but this; every man in the world could probably get a cartload of his own of corals here. The beach was undulating, like a ploughed field, with ridges parallel to the water, gradually increasing in height towards the land; all were quite bare and consisted of coral rubble stones, but the more one went inland the more they were covered by some mould. Here we had clear evidence of the annual increase of the land, for these corals or *Madreporae,* which cannot live anywhere except at the bottom of the sea, are thrown up on the shore, thereby enlarging the land. It could also be seen how late the stones had been covered with mould, since the ridges nearest the sea were quite bare, while those far away from the shore could hardly be recognized and those in between varied in proportion. The *Madreporae* lying close to the water were clean and clear, adorned with stars like the backs of playing cards or the cells of a honeycomb. Every stone star had a hollow cylinder in the middle, around which were up to 19-20 small squares. The cavities were filled with finely ground lime. When the stones were broken perpendicularly, the profile looked like a net with its *lamellis perpendicularibus* and *transversalibus decussantibus.* We also saw other *Madreporae simplices,* which looked like a finger-long horn of a calf, and had one single star at the thick end. Others looked like small goblets; still others were many times *proliferae e centro* like *Polytrichum* or Hair-moss, with one goblet inside the other. All these coral stones are, according to the findings of the learned botanist Bernard Jussieu, nothing but the shells of small worms, who make the stars and stone crusts, and there are as many species of these worms or *Medusae, Hydrae* and *Polypi* as there are varieties of *Madreporae.* All the creatures that build our corals remain to be described, and I have to leave this investigation to others, who have more time and choose the right season and weather for collecting them from the bottom of the bay. The further we came from the sea, the less we saw of the *Madreporae*, since, although every stone had originally been a *Madrepora*, it now looked like any ordinary limestone; sand and gravel stick to these corals and are incorporated in them, as if every *Madrepora* had the capacity of making stone; but time will show whether this is the way limestone rock is built. From the *Madreporae* sometimes a good and white lime was burnt.

Lepidium foliis pinnatis integerrimis, It. Oel. 52 [*L. petraeum* L. = *Hornungia petraea* (L.) Reichb.; see entry of June 2], was growing here on the coral shore.

Cucubalus: Lychnis maritima repens C.B. [*Cucubalus* 360 *Behen β maritimus* of Index = *Silene maritima* With.] was common on the coral ridges. Though the leaves were very small and the flowers were solitary, it was nothing but a variety of *Behen album* [*Silene vulgaris* (Moench) Garcke] or Campion.

Turritis [*Arabis hirsuta* var. *glaberrima* Wahlenb.] was also growing among the stones with smooth and shiny leaves. It was in all probability *Turritis foliis omnibus dentatis hispidis, caulinis amplexicaulibus* [*Arabis hirsuta* (L.) Scop.].

Draba foliis caulinis numerosis incanis, siliculis obliquis [*D.* 525 *gottlandica* of Index = *D. incana* L.] grew abundantly here, with its *folia radicularia* quite dense and close like a little *Sempervivum*. Its *caules* were *simplices, digitales: foliis versus radicem sessilibus, cordatis, acutis, utrinque denticulo notatis; Flores albi alyssi, sed filamentis denticulo destitutis. Siliculae ovato-lanceolatae, erectae, obliquae, pedunculis strictissimis; tota planta incana.*

Draba caule ramoso, foliis cordatis amplexicaulibus dentatis [*D.* 569 *muralis* of Index = *D. muralis* L.]. *Radix annua. Caulis filiformis, semipedalis, non tomentosus; Folia nulla ad radicem in cespitem digesta, sed caulina 7 vel 8 valde remota, cordata sessilia, amplexicaulia, serrata. Rami breves, laterales, ex alis foliorum solitarii. Racemus terminatrix, longus, laxus: pedunculis partialibus, horizontalibus, laxis. Capsulae ovales, parum obliquae, praecedentis dimidio breviores.* This one was growing in the meadows not far from the aforementioned shores and has never before been seen in Sweden.

Brunella [*Prunella vulgaris* L.] was growing here, too, with such extraordinary flowers that at first we thought we had discovered a new species, but we soon found out that it was only a variety [probably *P. grandiflora* (L.) Scholler].

Fläringe church, in the parish of Rute, was passed ½ a mile from Kappelshamn and the whole district between them with its bare ridges had the same kind of appearance. The church itself stood on a similar but smoother field, sloping down towards the sea and all bare, which together with the farmers' stone houses reminded one of Friesland.

The meadows were full of ramsons [*Allium ursinum* L.] and chives or *Porro capitulo bulboso. It. Oel. 60* [*Allium scorodoprasum* L.; see entry of June 4].

From Fläringe the journey continued to Hau, which was ¾ mile; we had to walk most of the way, for fear of breaking arms and legs if the horses should fall; the road was made of limestone shingle, so that one rode on it at the peril of one's life.

Globularia [*G. vulgaris* L.], which has not been found in Sweden except the one we found on Öland (*It. Oel. 65* [see entry of June 4]), was the commonest plant here; now that it had passed blooming, it was very much like *Jasione* [*J. montana* L.].

Red-brown sandy soil was found here, but it was so coarse that it could hardly be of any use at all. White bog lime was found everywhere around the limestone rocks.

Hau was a farm, at which we arrived at 8 p.m., and I must say this was the finest farm I saw in the whole country; there were no neighbours within ½ a mile; on two sides it borders on the sea and a little lake, and on the other two it is surrounded by vast sterile limestone rocks. Two tenant farmers lived here; each of them had his own white and beautiful stone house, with a farmstead of tarred wood. Everything inside the house was clean and proper; the kitchen full of copper kettles, large and small, 10-15 in each place. The farmsteads were surrounded by hop yards and gardens and huge leafy maple trees on which there were several small hollow wooden cylinders, wherein starlings and other small birds could lay their eggs and delight the inhabitants with continuous music. There were also beautiful wood piles to enclose the yard further. Here the well-known verse has come true: "En åtta kos bonde, som haver en häst, bor långt upp i skogen och fri för mang gäst, han mar aldrabäst." [A farmer with eight cows and a horse, who lives in the forest, free from many a guest, he of all lives best.]

We spent the night here at Hau.

June 28 [p. 194]

From Hau we went on to Rute church, but before we started we noted the following.

The "Allwar-lök" (*It. Oel. 53*) [*Allium schoenoprasum* var. *alvarense*; see entry of June 2] made the most impoverished meadows all red with abundant flowers.

Anthericum foliis planis corollis deciduis [*A. ramosum* L.; see entry of June 26] also grew here and bedecked the harsh meadows under which the limestone lay.

Anthericum foliis ensiformibus, perianthis trilobis, filamentis glabris [*A.* 269 *calyciflorum* of Index = *A. calyculatum* L. = *Tofieldia calyculata* (L.) Wahlenb.], a plant which I have previously found only in Lapland and described in *Flora Lapp. n. 137, t.10f.3.*, occurred abundantly in the same meadow, close to the marshland; this Gotlander was bigger in stem and flower and had a more pointed seed-vessel, but, like the Lapland one, was trigynous. This plant had a remarkable feature in that it has a three-cleft flower-covering, although no other of its kind or of its group has any such flower-covering; this had been already noted on this plant before me by the learned botanist J. Fr. Gronovius, who had received it from Virginia in North America.

Wood-dust which we saw in great abundance on some pine-trunks between the bark and the wood, was made by very numerous small white grubs that were feeding on the wood, and creating strange tunnels. These larvae were snow white, with a yellow-brown head, and were twice as big as their parents, but with lime their size is halved. The pupae were small and white, their heads also; they were as intolerant of sunlight as the larvae and they jerked themselves hither and thither when the bark was removed. The fully grown insect was *testaceum & undique pubescens, cujus coleoptrae undique punctis striatae, posterius quasi coronam obliquam ex denticulis fuscis; oculi oblongi obscuri.* Although this insect belongs to the same species as *Dermestes testaceus pilosus*

&c. [*D. typographus* L. = *Ips typographus* (L.)] that we have described earlier (*It. Oel. 26*) [entry for May 25] it was much bigger.

Rubia cynanchica [*Galium triandrum* Hylander] was growing everywhere along the road to Rute, with its three-cleft, more seldom four-cleft, white flowers.

After 1 mile we reached Rute church, where we botanized, while the theologians gathered in the church after the service to discuss something concerning them. During our "inquisition" of the pastures, we collected the following plants: *Gentiana corolla hypocrateriformi tubo villis clauso, calycis foliis alternis majoribus Fl. Lapp. 94* [*G.* 203 *Gentianella* of Index = *G. campestris* L. = *Gentianella campestris* (L.) Börner], *Carex spicis ovatis pendulis: masculina longiore erectiore, caule repente Hort. cliff: 439* [*C.* 762 *paludosa* of Index = *C. limosa* L.], *Brunella* [*B.* 498 *vulgaris* of Index = *Prunella vulgaris* L.] and *Astragalus Rivini* [*A.* 591 *procumbens* of Index = *A. glycyphyllos* L.].

Campanula caule angulato, floribus sessilibus, capitulis terminatricibus [*C.* 182 *glomerata* of Index = *C. glomerata* L.] grew in the pastures; its *Folia* were *cordato-lanceolata, crenata, pubescentia, petiolata; summa vero sessilia, semiamplexicaulia. Flores sessiles, in capitulum congesti, erecti, foliolis cordatis calycem fingentibus excepti; calyces simplicissimi.*

Serapias quae Helleborine latifolia montana C.B. [*Serapias* 734 *Helleborine* of Index = *Epipactis helleborine* (L.) Crantz] had *folia ovato-lanceolata* and bent the spike downwards so long as the flowers were not opened.

We passed Bunge church (in the parish of Rute) on our right, ¾ of a mile from Rute.

The meadows, which lay on the right-hand side after we had passed Bunge, had their fences planted like hedges with ash, mountain ash, blackthorn, whitebeam and hawthorn, that one could penetrate only with difficulty; thus has nature anticipated the art of growing hedges, which would be easier to grow in this district than in any other province of Sweden.

Whitebeam was growing wild in the pastures, in greater number than anywhere else.

We heard a bird screeching with an exceptional and strong voice, as if it were in a hawk's claws; but when we found it among the trees, it was nothing but a little chaffinch which had a big white butterfly in its mouth and was calling to its young for a meal.

From Bunge we travelled to Fårö; we had to go ⅛ mile to reach the sound.

Fårösund [Fårö sound], that separates Fårö from the rest of Gotland, is about ⅛ of a mile broad; but the length of the sound from the sea on the east side to the west side is about 1 mile; the mouth of the north-west side was very narrow and shallow, so that only vessels that draw 12 feet or less can enter; the eastern inlet is rather deep, and there are two islets there; it is thus a beautiful harbour.

Fårö is an island, part of a Gotland parish, which lies at the northern corner of Gotland, cut off by the already mentioned sound.

As soon as we had disembarked from the ferry boat, we shot a tern [*Sterna hirundo* L., common tern] which was sitting with its mate on the shore; one observed with sadness the affection of the free mate for the unlucky one; it flew above the other and would not abandon it when we took it up. *Atrum erat caput superne, incanae vero dorsum & alae supra; dilutius adhuc incanum pectus; Alba demum abdomen, cauda & alae subtus. Cauda forcipata: Rectricibus 1.1; 2.2. exteriori margine superne fuscis, & longioribus acutioribus, sed extima longissima. Remiges ab apice alae quo remotiores, eo etiam sensim breviores. Rostrum subulatum, versus apicem compressum, rectum, coccineum; Nares oblongae, lineares; Femora seminuda. Pedes coccinei, palmati. Affinitas summa cum Haematopo.*

The farmers here were clad in sweaters, nicely decorated in white, blue and red, which looked as if they were woven, but they were hand-knitted by the women.

Sandö lies 5 miles out in the sea from Fårö and belongs to Fårö parish.

Seal hunting is done on Sandö by half of the inhabitants of Fårö one year, the other half next year. The seal lard is sold for a price of 24 "styver" for a "lispund" and one seal yields about 16 lispund of lard. The seal meat is eaten fresh, salted or dried in the oven. The fresh fat is used instead of butter for frying pancakes. Seal calves are considered a delicacy. Among these seal-eaters we saw no contagious itches or rashes.

Porpoises [*Delphinus phocoena* L. = *Phocoena phocoena* (L.)], which are so common in the Baltic, are not hunted here or on Gotland, although their lard could be used as well as that of the seal. It is quite remarkable that nobody has developed the art of trapping them and that they alone should be left undisturbed. Sometimes, however, they are caught by accident in the nets.

The eider [*Somateria mollissima* (L.)] is common on the islets, but not rightly treated on the islands; the birds are shot and, what is still worse, their eggs are used for making pancakes. Indeed, throughout the spring one can see these birds decorating the fish market in Stockholm. The time will probably come when the excellent down of these birds will save them from being shot.

The tar that is distilled here is clear and brown, namely the true Gotland tar, made of clean roots, after the bark has been peeled off leaving only that part from which a transparent resin is obtained. These roots have often been lying in the earth for some hundred years, where they are found with difficulty and unearthed, and the pits from which they have been taken can be seen everywhere in the sandy forest; this is the special Gotland tar (the ordinary brown one is also made here) which is sold mostly to Germany and retailed over a large part of Germany. The inhabitants of this district have told me that the Germans use it for greasing their sheep once a year, which is supposed to make the wool softer and better.

We reached Fårö church at 9 p.m. at a distance of ½ a mile from the sound, and rested there during the night.

June 29 [p. 199]

Farö church was visited in the morning; on the wall by the altar was a remarkable event illustrated by a picture of 15 men and a lot of verses, which related how in 1603 the 15 men had gone out on the ice in the spring to hunt seals, the ice had broken away from the shore, and all the men on the ice raft were forced to go to sea, and as they passed Sandö, three of them managed to jump to other ice floes and thus reach Sandö, the others had to follow the ice and the wind to the wild sea, but, through the protection of the Almighty, they all arrived alive in the Stockholm archipelago after a voyage of 14 days, with nothing to eat but raw seal meat.

The following verses are written in the Danish language on the painting:

We, men of Fårö, rightly now may tell
Of our fate to rich and poor alike.

Behold this painting, read and mark well
How mighty is God, even in great adversity.

We have caused this record to be made
That for others the tale may be retold.

In the year sixteen hundred and three
On the day of Saint Matthew, in the thaw
Out upon the ice we ventured—
To hunt the seal was our desire.
But our fortunes took another turn,
Our fate was changed by Spring and sea.
Right fifteen men we were in all
Borne out to sea upon a floe of ice.

Jacob Nors I should mention first,
Christopher and Eschil and Nors Michel
Thomas Butley and Anders Oesters
Staphen Gussemor and Hans were there,
Rasmund Simunds and Hans Simuds
Rasmus Ringvid and Peder Sudergard
Bottel Södergard, Jurgen Mor, Hans Mor
were out upon the ice cut off from land.

We were borne from Gotland and Fårö isle
Speedily away t'wards Gotland's Sandö
Where three of the men we named before
Left us to try for Sandö's strand—
These were Peder and Bottel Södergard
and Jurgen Mor made the third.
Providence brought them home again
To Fårö isle both safe and sound.

The rest of us upon the ice remained
We must needs follow it we knew not where
In anguish, hunger, frost and great distress
Faced always with a grim and bitter death
Our only fare, the uncooked meat of seal,
We had to eat without a crust of bread.
For fourteen days did this sustain our lives
But then came God with timely help
So that we drifted in to Sweden's land,
And were released from the agony of death.

Our Lady's Day in Lent was the day of our return
To our Fårö isle, to wives and children, house and home.

There were some runic stones in the churchyard; one of them was impossible to read, but the other was easier, since a fine *Lichen crustaceus leprosus* [*Lecanora calcarea* (L.) Sommerf.] grew only in the letters, making them visible at a long distance, but when the lichen was rubbed off, the letter became quite illegible too.

Early in the morning we left Fårö church and rode along the eastern shore towards the northernmost point of Fårö, 1¼ miles from the church, and returned the same distance along the western shore.

Four kinds of seaweed are used as manure: 'hauter'', "tang", "kräkel" and "ylle", all of different value. "Ylle" (*Conferva marina ramosissima*) [*Pilayella littoralis* (L.) Kjellm.] is the best manure, but when too much of it is spread on the field it does harm.

"Kräkel" (*Fucus dichotomus ramosissimus teres uniformis fastigiatus*) [*Furcellaria fastigiata* (L.) Lamour.] is not as good as "hauter" or "ylle", but somewhat better than "tang".

"Tang" (*Alga angustifolia Vitrarium I.B.*) [*Zostera marina* L.] is the same as "gräsja" of the Ölanders; this is regarded as the least useful.

"Hauter" (*Fucus folio dichotomo integro, caule medium folium transcurrente, vesiculis verrucosis terminatricibus*) [*F.* 1102 *Quercus marina* of Index = *F. vesiculosus* L.] the Gotlanders and Ölanders call "slake"; it is especially good after having been heaped and become well rotted. We saw huge stacks of it in which the cattle were prodding, doubtless to lick the salt. The cattle were said to eat a little of it during the spring, when it is still fresh and recently taken from the sea. "Haute" [bladder wrack] is also used in charcoal stacks to damp and choke the fire, instead of earth.

The corn fields looked like banks or shores of white stones, for hardly any black soil at all was seen, except for the bladder wrack that was mixed with the stone; nevertheless the rye was growing beautifully, and the only disadvantage of so much stone was that the farmers have to use twice the normal amount of seed for sowing, since much of the grain is choked by the stones.

Fårö, as much as we travelled over today, was sterile land under which limestone lay, but nearer the sea it consisted of sand and stone. It is not as flat as Öland, but there are no rocks, hills or deep valleys. The woods consist of pine trees and junipers, which grow unevenly and have been felled extensively, but at the north corner of the island they were much better. Deciduous trees were uncommon in the pastures; the tall pines looked ugly. Hazel was here fairly common, but not as much as in Öland. Fir trees were rare on this island.

We reckoned 18 lakes on Fårö which, although they were small, were all rich in fish, and this by no means includes any small puddles, which dry up during the dog-days.

The grain that is grown here is mostly rye, since barley does not readily grow in these dry fields; the rye yields a white bread and usually enough grain for the farmer, who can also earn some money from herring, cod, flounders, seal-lard and tar.

Pastures are common all over the island except for some small enclosed ones, but meadows and corn fields are fenced.

The houses and outbuildings are here well built. The roofs are made of double boarding only, but the barns are usually thatched with sedge, twice as thick as a straw roof. There are often some small stone houses on the farms.

The stone houses are built of limestone, sometimes joined with mortar, sometimes with bog lime, clay and sand, although the bog lime does not bind as well as proper mortar. These houses are whitewashed on the outside and roofed with tiles, which is no little embellishment to the countryside; one can see that this manner of building stone houses on Fårö is not ancient.

Ave farm was situated ¾ of a mile from the church on the eastern shore; it looked neat and tidy; here we cooled down from the burning heat under a huge oak which stood in the farmyard.

The oak was one of the most considerable in size; its trunk was 7 ells [14 feet] in circumference, its height was 37 ells [74 feet] and the spread of the branches, i.e. the diameter of the crown, was 44 ells [88 feet].

A net for seal-catching made of treble twine was hanging here on the fence; every mesh was one "quarter" [6 inches, 15 cm] square, tarred and black; the edging was as thick as a goose pinion, made of black horsehair, and was sewn to the net with a horsehair thread; small wooden floats were attached to the edging, each a foot [12 inches, 30 cm] long and a few inches broad; the height of the net was 3 ells [6 feet], that is 15 meshes, the length 8 fathoms [48 feet], with a big stone tied to one end of it, so that the seal cannot escape with the net; as soon as the seal is caught in the net, it is killed by the seal-hunter.

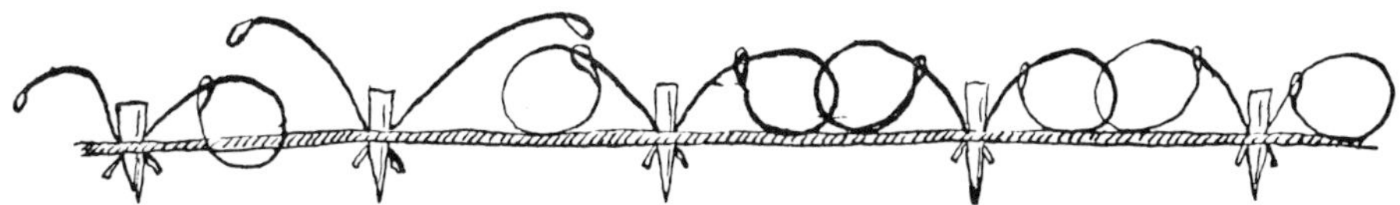

Bird snares to catch sea-birds were lying on the shore; the line was quite long; at every half ell [12 inches] was a wooden peg stuck into the ground to fasten the line; on both sides of each peg there is a snare of black horsehair, 3½ "quarters" [21 inches] long, which snare is strengthened at the base with a pinion to make it stiffer. The line is placed far out along the beach, so that ducks and other small sea-birds are trapped when they leave the water.

Chrysomela viridi-aenea [identity uncertain] with one and the same colour over the whole body, and the size of a ladybird with 7 black spots or somewhat smaller, was found in the wood.

Dunes of shifting sands lay here and there at the shore, thrown up high by the sea, but at the northernmost point the whole land was covered with this sand forming high and uneven sand-hills. This reminded us of the Dutch dunes. The sand in these dunes was fine, white and clear. The irregularities in the sand hillocks were caused by the "sandhafre" [marram grass] which grew here and retained the sand. Within these sand-dunes one saw sand-ridges or embankments of the same kind one after another in the land, but the further inland the lower they were. Stagnant water separated these sand-ridges like ditches; here grew *Drosera foliis orbiculatis & oblongis* [*D. rotundifolia* L., *D. longifolia* L.], *Potamogeton foliis natantibus* [*P. natans* L.], *Chara vulgaris, Juncus culmo nudo stricto, capitulo laterali* [*J. conglomeratus* L.] and *Centaurium minus offic.* [*C. erythraea* Rafn].

The "sandhafre" (*It. Oel. 139* [entry of June 18; *Ammophila arenaria* (L.) Link]) or the Dutch "hälm" grew everywhere here among the sand hillocks. A strange grass, which in the driest sand grows rampantly both high and tall, and the further it comes up in the sand the more runners branch off, one from each joint; the leaves under the sand wither and become dry, which makes the lower part of the plant look like a broom, from which the sand cannot escape. We dug to try to find the grass's lowermost root, but we could not dig deep enough in the loose sand; we noticed, though, that the grass grows not only straight up in the sand, but also out sideways. Wherever marram grass is lacking among the dunes, the wind blows the sand away and leaves a hollow, and thus it comes that the hillocks are so uneven. The use of this grass is varied, since where it grows the shifting sands cast up by the sea are retained on the shore and cannot drift inland and cause harm; the more sand that drifts on the grass, the better it grows and the higher the hill becomes, hence one saw smaller sand hillocks, looking more like pasture land, between the mainland and the shore, since the larger hillocks closer to the sea had taken their supply of sand; on the smaller hillocks the marram plants were stunted, being deprived of fresh sand, on which they can grow luxuriantly. Thus the land increases through the action of the marram grass, since the sea heaves up fresh sand on the beach each day; but as it cannot pass the dunes, it stays outside them, and the roots of the grass creep into the sand, fasten it and make the dunes bigger towards the sea, whereby the land grows. The marram grass with its sand-hills also prevents the sea from flooding the land during the winter; in Holland it can be seen how the dunes between Haarlem and the Hague form a barrier that shuts out the water from the low-lying Netherlands.

Formica-Leo was found among the dunes, far more multicoloured than on Öland (*It. Oel. 149* [entry of June 19]); we noticed what has later been written about it by the great entomologist Réaumur in his *Histoires des Insectes, T. VI. Mem. X. Tab. 32, 33, 34.*

Salix foliis integerrimis ovatis acutis: supra subvillosis subtus tomentosis [*S.* 806 *sericea* of Index = *S. arenaria* L. = *S. repens* var. *nitida* (Ser.) Wender.] is a small willow bush, which grows everywhere on the inner and older dunes; the stems were not much longer than blueberry sprigs, the leaves were ovate, pointed, not toothed at the margins, but recurved, downy on both sides, white, and glossy like silk; the petioles were so short that they could hardly be seen; the shoots were reddish, but the youngest ones pale and downy.

Pyrola floribus undique racemosis dispersis, staminibus pistillisque rectis [*P.* 331 *Halleri* of Index = *P. minor* L.] is a plant that I had not had the opportunity of seeing before; it grew among the older dunes; it was as like the *Pyrola* sold in the pharmacies as are two eggs in a basket. The leaves were ovate, the low ones smaller, the upper ones more obtuse (*retusa*); the inflorescence was like that of the common *Pyrola,* but the flowers were more drooping. Stamens as well as the styles were straight and protruding, not curved as in the common *Pyrola.*

Red colouring was found in the dune slacks, like red ochre or the "byttelet" in Västergötland; it was the same colour as we found on the sea-weed at Kläppinge in Öland (*It. Oel. 110* [entry of June 11]), perhaps produced by sea-weed, rotted a long time ago.

Carabus nigricans pedibus tibiisque pallidis [*C. velox* L.* = *Bembidium velox*

* The "arena riparia maritima insulae Färoen in Gotlandia" is the type locality for *Carabus velox* L.; cf. L., *Fauna Suec.* 2nd ed., 222 no. 803 (1761).

(L.)], an insect that ran on the beach in great abundance; it was so agile that, although it had shells [i.e. elytra] on its wings, it was swifter in flight than any bee; it was completely black, as big as a housefly, and the "wing shells" [elytra] were grooved with 8 fine lines; the "breast-shield" [thoracic cuticle] was smooth and the "horns" [antennae] half as long as the body.

The most northerly point of Fårö had nothing remarkable about it but a flat sandy beach, on which the sea had sketched several mountains and high alps as with a pen.

Ducks in millions, all quite grey, were swimming in the water and flying about us. The many sea-birds that hatch their eggs in Sweden mostly leave here uninjured, but they are caught in Holland and other southern countries, so that the market-places are crowded with them; our nation has not yet reached such perfection in bird-catching.

"Tang" (*Alga*) [*Zostera marina* L.] was lying on the shore looking like thin black tape-worms; some were pitch dark and opaque, others snow white and translucent.

Conferva ramosissima capillacea, pallida, internodiis oblongis was also thrown up on the beach.

Serpyllum vulgare minus [*Thymus serpyllum* L.] grew here in the sand, but quite rough and hairy.

The journey back was along the western shore.

The terns were flying in such numbers that we were able to lash one with a whip when it was flying; it fell to the ground, half dead; it was similar to the one we shot yesterday at the Sound, but the beak was black in the upper part up to the tip.

There were two sand islets, one outside the other, beyond the northernmost point, on the western side; the farmers told us that it was possible to wade out to them on horseback when the water was low; these islets will in all probability be incorporated with the mainland in the future; fine crops and grass were growing on them now.

The meadows on former corn fields were yellow with *Ranunculus acris* and *Lotus corniculata glabra* [*L. corniculatus* L.].

The Ölanders' "Gräslök" (*It. Oel. 60*) [entry of June 4] is called "Kaipe" by the farmers, a word that seems to be derived from καιπη or *Cepa*.

Bog lime was seen everywhere; we asked the farmers if they ever use it as manure, since limestone is said to fertilize the soil, but they know nothing thereof.

"Butta" is the local name for a flounder [*Pleuronectes flesus* L. = *Platichthys flesus* (L.)] with eyes on the grey and unspotted left side; the jaws were full of small teeth, and the gill-cover (*membrana branchiostega*) had 7 rays, not 6. The rays were 64 in the dorsal fin, 11 in the pectoral fin, 6 in the pelvic fin, 48 in the anal fin and 16 in the tail fin. The smaller type of flounder with bright orange spots (see entry of June 27) varied with the eyes on either the left or the right side.

Local household remedies, which are rather rare in the country, we learned of from the farmers. "Hansletgräs" is the name for *Solanum officinarium C.B.* [*S.* 188 *officinarum* of the Index = *S. nigrum* L.], which herb is pounded between a couple of stones while it is still fresh, together with spider's web and rancid pork, and put on whitlows (*Paronychia*), which are called Hanslet here;

they assured us that this was a sure remedy, provided the disease was not too old. Another cure for this disease was an egg yolk and the same amount of salt, well mixed, put on the finger.

Ballota [*B.* 484 *vulgaris* of Index = *B. nigra* L.] is boiled and given as a *panacea* to the cattle when they are sick.

The following dye-plants are used: *Lichen fulvus sinibus daedaleis laciniatus* [*Cetraria juniperina* (L.) Ach.], which grew on old wooden fences and junipers, yields a yellow dye when cooked with alum. Stone moss, so often used for dyeing, was also known by the farmers as growing on granite and not on limestone. Yellow-bark (*Cortex Frangulae*) [bark of *Rhamnus frangula* L. = *Frangula alnus* Miller] is boiled in fresh water, without salt, then yields a yellow colour, but if it is first dried it dyes brown; the cloth is boiled together with the bark. Goat-bark (*Cortex Rhamni cathartici*) [bark of *Rhamnus catharticus* L.] is boiled in water, and the decoction is spread on the cloth and dried in the sun, which gives a dark brown colour. The nets are dyed red with birch bark, cooked in a lye made from hazel ashes.

"Linbär" [flax-berries] is the local name for bearberries, which occupy the whole country.

"Tulke-Gräs" is the name for *Asclepias* [*A. vincetoxicum* L. = *Vincetoxicum hirundinaria* Medicus], which grows here over whole fields.

The language here on Fårö is somewhat more difficult to understand than that of the rest of Gotland; the farmers talk in a broad dialect and change their vowels into diphthongs, such as "jau" for "ja" [yes] and "naj" for "nej" [no]. They had also many words that were altogether different from ordinary Swedish, such as Häst, *russ, pert;* Vrensk Häst, *fast*; Bandet över myssan, *lausholk*; Spegal, *söndags-au*; Livstycke, *baudordi*; Gosse, *sork*; Spener, *mammer*; Piga, *pika*; Litet stycke, *lillebisken*; Mossa, *muus*; Sadel, *fis-flacko.*

We lodged the night once more at Fårö church.

June 30 [p. 210]

We crossed over the sound once more at the same place. At Fårösund the farmer-folk were having a feast.

The feast included dancing by farmers and their womenfolk. The measure of the dance was very strange, *ex uno pede longo, altero duplici saltu brevi.* We were regaled with "lura", a drink that is opaque, like the colour of white wax, and boiled by [dropping in] glowing grey-stones. This drink had been brought by a man from Ösel, who was also playing a bagpipe for the guests.

The bagpipe was remarkable in that it had no seams; it was made from the whole stomach of a seal; one blew into it by *fundo ventriculi versus pylorum*; the modulator was placed in the *pylorus* and the base, which was hanging down, was inserted in the *oesophagus.*

Smoke-pits for smoking flounder and herring were built into the earth and sealed so tight that the smoke could not escape; they were black on the inside, 3 ells [6 feet] long, 2 ells [4 feet] broad and 2 ells [4 feet] high, and only the roof was visible above earth. The fish was hung up and smoked with fir and pine cones or even rotten oak or decayed pine stubs as fuel, or anything that would burn without a flame, which smoking lasted for 4-5 hours only, so that the fish would not become too dry.

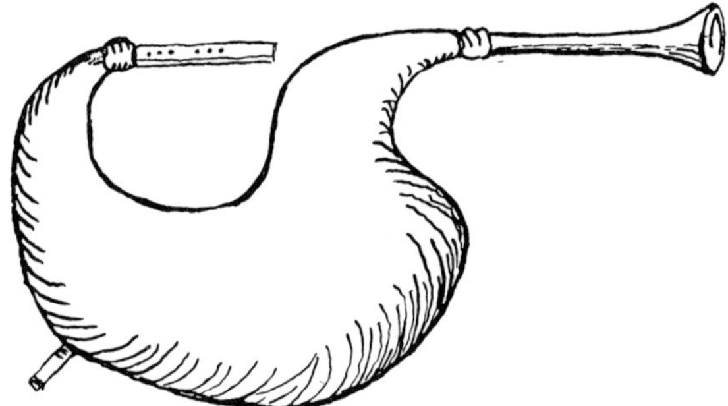

Coral, especially *Milleporae* and *Celleporae,* of which the edge looks like a chain, were found in great numbers along the shores at Fårösund.

Bunge church, in the parish of Rute, was passed once more; it stands on the top of a hill from which the land slopes down towards the sea on the east side; from an architectural point of view, this was the most splendid church we had seen so far on Gotland.

The farmers told us about some rocks to the east not far from the church, of a kind unknown here, which undoubtedly contained some metal; Herr Adlerheim, who rode there, returned with a couple of these stones, of which one could say that they were *Spatium incarnatum lamellatum cum mica membranacea fissili alba.*

From Bunge we went southwards along the eastern shore through the woods, which had been considerably thinned out by the lime burners, but here and there were some fir trees.

The tar kilns were here built above earth and look like inverted cones with the floor of stone and the many-angled walls of timber.

The plants seen here were *Trifolium lupulinum* [*T. campestre* Schreber] in great amounts in the forest, *Behen album* [*Silene vulgaris* (Moench) Garcke] in the pastures. *Buphthalmum luteum* [*Buphthalmum* 698 *campestre* of Index = *Anthemis tinctoria* L.], *Verbascum, quod Tapsus barbatus* [*V. thapsus* L.], *Jasione* [*J. montana* L.] and *Trifolium quod Lagopus* [*T. arvense* L.] were growing here and there.

Scutellaria foliis integerrimis: inferioribus hastatis, superioribus sagittatis [*S.* 502 *integrifolia* of Index = *S. hastifolia* L.], was found on the seashore. This plant, although rather well illustrated by Rivinus, has been regarded by botanists as a variety of the common *Scutellaria* [*S. galericulata* L.], from which it is quite distinct. The root creeps. The stem is as high as a finger, without branches, square, with six joints. The spike is as long as the stem, loose, blue, one-sided, so that the flowers arise like the leaves in pairs, 2 or 3 at each joint of the spike. The leaves are like *Linaria Elatine dicta* (*hastata vel sagittata & auriculata simul*) [*Kickxia elatine* (L.) Dumort.], especially the lowest, which have a longer stalk, but the other leaves are arrow-like and attached to the stem almost without a stalk; the leaves in the flower spike are ovate-lanceolate; all leaves are entire at the margin. The calyx has a shield on the back. The lower lip of the corolla is wavy, but not notched. Some of the plants we found had branches from the root [see p. 12].

At the Malms farm we saw eiderdown, as it looks when in the bird's nest; the down was grey with white spots and so tightly stuck together that one could shake it without the down flying about; the down was very soft and swelled out again by itself after being squeezed together. The birds pluck this down for the nest to cover the eggs and keep them warm, when they leave the nest to find food. The nests are found in the district especially on Furillen island and generally under the bushes. The down has in it much moss and small twigs which are difficult to remove; this is done with a taut piece of twine, about a fathom [6 feet] long, with which the down is spread in much the same way as a hatmaker spreads his wool; whereby the clean down remains hanging on the twine. One "mark" [pound] of eiderdown is sold for 10 *Caroliner in specie.* Four "marks" of down are needed for making a small bed, which can be squeezed to the size of a man's head but spread out to the size of a bed when released, which makes it valuable both for its lightness and its warmth. The eiderdown is not spoilt if the birds are allowed to hatch the eggs, but is left clean in the nest when the young birds leave it; therefore both eggs and birds should be more privileged than they are. The eider-duck tastes of whale oil, but the inhabitants remove this taste by parboiling the bird with hay before it is roasted.

"Ju-gäs" [*Tadorna tadorna* (L.), shelduck] is the local name for a sea-bird that is shaped like a goose, but somewhat smaller; it has down almost as fine as that of the eider, but is all white. We only saw a female, which was white, but with breast above rusty red; abdomen was lengthwise spotted grey; head and throat above greyish; pinions were rusty-coloured at the tip, so that a large rusty-red spot appears on the side of the wing. The males are said to be of another colour and to have a *protuberantia carnosa* on the bill, of which we saw no sign on the female, whose bill was very like that of a goose but more depressed in the middle.

"Grylle" [*Uria grylle* (L.), black guillemot] was a sea-bird likewise here in the sea; it was smaller than a hen, quite black except for the red webbed feet which consist of 3 toes without a thumb [hind claw]; the pinions are also black; but the wing-coverts are white, which gives it a big white spot on the side of the wing; the tail is short. The bill is awl-like, almost like that of a hen, convex and black; the upper jaw is longer than the lower, and there are no teeth at the margins of the beak. This bird is a kind of *Colymbus* or diver.

A gull [*Larus marinus* L., greater black-backed gull] was shot here, which was quite big and all white except for back and wings that were dark; the pinions were all black except for the white tips which were, however, hardly visible except on the first pinions. The first pinion had a large white spot at the tip. The thighs were half naked, feet yellow, webbed. The bill was pale with oblong nostrils, the lower jaw had a scarlet red knob.

Kyllej was our resting place for the night, after we travelled 2¼ miles in intense heat and sweat.

July 1 [p. 215]

We visited Furillen, the island near Kyllej, early in the morning, but it had been grazed and there was nothing remarkable to see except *Scutellaria foliis integerrimis* [*S. hastifolia* L.], *Spergula foliis oppositis, pedunculis simplicibus*

[*S.* 378 *nodosa* of Index = *Sagina nodosa* (L.) Fenzl], *Prenanthes* [*Lactuca muralis* (L.) Gaertner = *Mycelis muralis* (L.) Dumort.] and *Veronica floribus spicatis, foliis oppositis, caule erecto* [*V.* 7 *spicata* of Index = *V. spicata* L.].

Chara caulium articulis inermibus diaphanis superne latioribus [*Ch.* 995 *glabra* of Index = *Ch. flexilis* L.] grew under water at the seashore; the leaves were placed here and there together like bristles, with small red seeds at their bases.

There were several caves in the precipitous rocks which lay a little back from the shore at Furillen, these caves were some 5-6 ells [10-12 feet] above sea level; earth and stones had been scoured out at the time when the sea reached this level with its shore. Now they were excellent shelters and houses, which nature had made for the grazing herds of cattle and sheep, which rested here at night and were protected from rain in the day-time.

"Swärta" [*Melanitta fusca* (L.), velvet scoter]. A female was shot, which was slightly bigger than a duck; the bill was similar to that of a goose, dark at the edge, with many 'teeth', like upright scales. The edge of the tongue had hairs looking like eyelashes. The whole bird was dark (*fusca*) with pale feather tips. The first 10 pinions were pitch dark; No. 11 was black with a white spot on both sides; Nos. 12-20 were white with almost black tips; No. 21 was like No. 11. The feet were red with a black membrane which joined the claws. A white spot was behind each eye, and the ear-coverts were also lighter.

Aranea abdomine antice lateribus acuminato [*A. angulata* L.] or a rather big spider was found in the trees; the abdomen went from the breast on each side into an obtuse angle and was variegated with multi-coloured white, yellow and black, but the legs were only white and black.

Klasen is the name of another island, not far from Kyllej; this island is rather long but not very broad. All of it is pasture. Here grew: *Scirpus culmo triquetro, panicula conglobata* [*S.* 39 *maritimus* of Index = *S. maritimus* L.]; *Veronica spicata minor C.B.* [*V. spicata* L.]; *Plantago angustifolia minor Tabern.* [*P. lanceolata* L.]; *Thalictum*; *Draba alpina hirsuta Celsii* [*D. incana* L.]; *Rhinanthus qui crista galli foemina J.B.* [*Rh. minor* L.]; *Melilotus officinarum* [*M. altissima* L.], that is sold in the pharmacies, also grew in great quantity.

Woad or *Isatis* (*543*) (*It. Oel. 144*) [entry of June 18], the useful dye-plant, that hitherto had been thought so rare in Sweden, grew on the south and west shores of Klasen where the sea had thrown gravel and sand together to form a bank 1 fathom [6 feet] high at 3-4 fathoms [18-24 feet] from the waterline; this area between the bank and the sea was covered with seaweed or *Fucus.* On this bank the woad grew like a field of hemp, and it could be no more luxuriant in a garden. It would be well worth while to sow this dye-plant all over the island, since it grows so well here.

Scabiosa corollulis quinquefidis, foliis radicalibus ovatis crenatis, caulinis pinnatis setaceis [*S.* 111 *tenuifolia* of Index = *S. columbaria* L.], a rare plant, that grows wild in Germany and sometimes adorns our gardens, but has never before been found wild in Sweden, was growing on the slopes of Klasen. *Folia radicalia latiuscula, duplicato-pinnata; superiora linearia, simpliciter pinnata; intermedia itidem simpliciter pinnata, sed pinnis linearibus extrorsum ramosus. Calyx in 15 vel 19 angusta segmenta divisus cingebat florem nondum explicatum.*

Birds such as oystercatchers (*Haematopi*), gulls (*Lari*) and "tolks" started to worry us and to fly around our heads with intolerable screeches, so that we heard nothing else after we had gone ashore. We shot a "tolk", a bird that we did not know, either in appearance, flight or call.

"Tolk" (*Tringa nigro albo ferruginoque variegata, pectore abdomineque albo*) [*Tringa interpres* L. = *Arenaria interpres* (L.); see p. 13] is a bird that I have not seen in any other place, nor in any author. It was like a lapwing, but not bigger than a thrush; abdomen, breast tail and wings on the underside were white; the head was white above with small yellow-brown spots; the throat below and part of the breast were covered with black, which sent out a black stripe on each side to the upper part of the throat and another black stripe from the eyes to the forehead and another one to the lower jaw; between these stripes were white spots, making a white patch under the bill, another at the eyes and another on the forehead. The tail was short and dark; all the tail feathers, except the two central ones, were white at the tip and at the base. The wings were grey outside; the pinions were blackish and like the tail feathers, namely, the outer ten with a white base, but the inner ten white both at base and tip. The back was covered with loose and long feathers, of which those on the sides were black with rusty red spots. The bill was short, awl-like, blunt and black, not longer than the toes, with oblong nostrils into which one could look. Thighs half naked; legs blood red; feet unwebbed with three toes pointing forwards, one backwards.

We called the "Stone Giants", as had the learned bishop G. Wallin, that which we saw by the sea near to Kyllej: between the customs officer's house and the lime kiln of Kyllej there was a slope towards the sea where stood many high and thick limestone rocks 4-6 fathoms [24-36 feet] high, arranged in a row like the ruins of churches or castles, of which those standing at a lower level of the slope were taller than those higher up, so that the heads were all at the same level. From a distance they looked like statues, horses, torsos and I do not know what kind of ghosts. Evidently, this had been formerly a limestone mountain, the roots of which had been ground, cut and formed by the heaving waves of the sea, till it finally left these stones in their present form; there is no doubt that the water that has cut into the sides of these cliffs and made them narrow at the base has also been able to cut away and erode the earth between them. Stone giants of the same size were seen all the way along the road to Slite [see Plate 7].

From Kyllej we rode southwards.

Hällvik church, in the parish of Lärbro, was on our right-hand side, and we turned off the main road on the left-hand side for St. Olof's church $^3/_8$ of a mile from Kyllej.

St. Olofsholmen was a round and hilly islet that has recently become attached to the mainland on two sides. On the part of the islet facing the sea stood part of a church 20 ells [40 feet] broad and 40 ells [80 feet] long, about which, and about St. Olof himself, we were told much; we were shown a big stone with a hollow in it, as big as the crown of a hat, which was called St. Olof's wash-basin.

Walnut trees had been planted on one side of the islet, but they had been destroyed during the last two hard winters; this harsh bare island so exposed to the winter storms is not a good place for this tree.

Plantago foliis lanceolatis longitudine scapi, spica oblonga [*P. lanceolata* L. var.] grew here, as similar to *Plantago major angustifolia* [*P. lanceolata* L.] as possible, regarding both the spike and the truncated root, that one could imagine that nature wanted to clothe it in long hair to protect it from the chilly sea-winds, as was also the case with *Serpyllum vulgare minus C.B.* [*Thymus serpyllum* L.], which had downy heads like marjoram, wherefore Tournefort calls it *Serpyllum vulgare minus capitulis tomentosis.*

The distance from St. Olof's chapel to Slite was 1½ miles. The woods today were quite dense, though not especially old, with more fir and more *Hypnum* than we had seen before; here grew *Brunella flore magno* [*Prunella grandiflora* (L.) L.] and my flower [*Linnaea borealis* L.]. Here and there between the fields were dry stone walls, on which *Ruta muraria officinalis* [*Asplenium ruta-muraria* L.] was growing.

Slite, one of the finest harbours on Gotland, provided us with a night's lodgings.

July 2 [p. 220]

Karlsvärd, a fortification founded by King Charles XI on a little island called Enholmen, could be seen from the mainland, but a strong east wind and lack of time prevented our going there; the scythe had begun to violate our plants, and typhus had afflicted two people in the house where we had stayed, as well as 15 people on the nearest farm.

Sju Strömmar was situated by a little lake, Bogevik, at its outlet into the sea almost ⅛ of a mile from Slite; the land between lake and sea was ½ a gunshot [about 120 yards] broad. Between lake and sea four channels had been dug, each a fathom [6 feet] deep and a few fathoms wide, though the water at this time of the year was hardly more than 1½ ells [3 feet] above the clean bottom. Sometimes water flows from the sea into the lake and sometimes fresh water from the lake flows into the sea, depending on the wind, as in Stockholm. When the fish in the sea notice the fresh water they are lured to swim into the lake through channels and are caught by the farmers with bag nets in a very simple manner; when the farmers see the fish in the channels they come at them from both inlets with bag nets. The main road crosses these four channels by as many bridges.

Boge church, in the parish of Othem, was seen on our right hand on the other side of the lake.

Tjällvards island and Tjällvards bay were passed on our way; this islet is not big, quite high, attached to the mainland, although it has formerly been an island proper. Tjällvards island is said to take its name from a man called Thielward who landed here before the birth of Christ and occupied Gotland, if the historians, with their good memory, tell the truth. In the meadow *Anthericum perianthis trilobis* [*Tofieldia calyculata* (L.) Wahlenb.] (June 28) and *Lotus leguminibus solitariis membranaceo quadrangularibus* (*It. Oel. 143*) [*L. maritimus* L.; see entry of June 18] were growing.

The woods, which we passed through, were rather impressive, consisting of fir trees, before we came to Gothum river, but on the south side of this there were mostly pines with tall trunks and the ground cleared of bushes, so nothing was missing to make good timber except time.

Gothum river was the biggest river in the country, but the water in it was now so shallow that we could wade over it in boots.

The plants around Gothum river were especially *Hedera* [*H. helix* L.], trailing on the ground under the bushes, *Pyrola scapo unifloro* [*P.* 334 *uniflora* of Index = *Moneses uniflora* (L.) A. Gray], *Pyrola racemis unilateralibus* [*P.* 332 *secundiflora* of Index = *Ramischia secunda* (L.) Garcke]. *Pyrola staminibus adscendentibus pistillis declinatis* [*P.* 330 *irregularis* of Index = *rotundifolia* L.] had a three-angled stalk like *Pyrola floribus undique racemosis* [*P.* 331 *Halleri* of Index = *P. minor* L.] (June 29) but the corolla was paler and more open. *Anthericum calycibus trilobis* [*Tofieldia calyculata* (L.) Wahlenb.]; *Linnaea* [*L. borealis* L.]; *Nymphaea alba* [*Nymphaea alba* L.] & *lutea* [*Nuphar lutea* (L.) Smith]; *Alisma* [*A. plantago-aquatica* L.]; *Chara caulibus aculeatis* [*Ch. hispida* L.]; *Potamogeton pusillum fluitans Bock* [probably *P. filiformis* L.], with a *caulis ramosissimus; foliis alternis, lineari-subulatis, angustissimis; ad ramificationes vero caulis notatis: stipulis latis, amplexicaulibus, e quarum dorso folia solitaria.*

Gothum was reached after a 1¼-mile journey; the vicarage seemed to have been a monastery to judge from its heavy walls and arches. In the dining-room arch there was a window like an eight-pointed star, on the outside little more than a quarter-ell [6 inches] big, but much broader in the inside, through which the sun shone at 10 a.m. The church was beautiful. Men and women here sang one verse each of the hymns, while the other sex kept quiet. A big runic stone with much lettering lay in the chancel.

Ash trees and elms grew by the fences, all of them lopped by saltpetre boilers, who told us that the ash of these trees was indispensable for removing scum from the lye, for they considered the ash of oak to be unsuitable for this purpose.

The goats are left in the open winter and summer and get no more from the farmer than what they steal from him.

Squirrels had been seen to sail over the lakes on chips of wood or bark according to the people here, and they are supposed to disappear every 7th year. Thereafter missed very much, they then increase in numbers until the 7th year comes again. This is quite a new tale to me, which should be investigated. Weasels are also found here.

Household remedies, which one learned from the peasants' wives, included pepper with strong liquor or hot beer and strong pepper (*Cubeber*) for after-pains. Tar was used by those afflicted with *siphilis*, which made the ulcers in *partibus genitalibus* heal; but their *praetumidae nymphae, Hippomane, Tophi, Dolores artuum nocturni, Ulcera Laryngis* and *obstructio nasi* showed clearly of how little use tar is for such a disease.

Yew trees grew as big as firs or oaks, mostly in the marshlands, in the parishes of Gothum and Boge. The inhabitants had a nice manner of covering their walls with yew twigs, starting from the floor and covering all the wall as with shingles, for which the soft needles made a delightful green wall cover. If *Dioscorides* and *Plinius* had been invited to a house with yew-covered walls, they would never have dared to sleep there one single night or to eat any food, since they believed that sleeping or eating under a yew tree was fatal; this makes people laugh here in Gothum.

"Schiäde" (*Lolium spicis aristatis, radice annua*) [*L.* 163 *fatuum* of

Index = *L. temulentum* L., darnel] was growing here abundantly and in many places among the barley, but never among the rye. Those who drank beer brewed from darnel-mixed barley became mad and almost blind. The farmers thought that if they rubbed their wrists and fingers with the beer, it would protect them from the affliction.

The clay, which comes from Gotland and is much esteemed by some people in Stockholm, was seen here at Westers farm along the Gothum river, ⅛ of a mile from the church; it was nothing but ordinary clay mixed with bog lime, which was no good for clay pipes, although the clay did turn white in the oven; it was much too brittle.

The dye-plants, which are used here in the parish, were as follows: *bright yellow* was dyed by means of "flowers of St. John", for which the yarn is boiled with alum in a copper kettle, although not with too much alum, then it is dried; after that dry "flowers of St. John" are boiled for a long time in water, and finally the mordanted yarn is added to the decoction. It should be noted that these "flowers of St. John" are not *Flores Hyperici* [flowers of *Hypericum perforatum* L.], which are usually called thus in Sweden, but are the *Flores Buphthalmi* [flowers of *Anthemis tinctoria*]. *Greenish yellow* is dyed with birch leaves boiled in water after mordanting with alum. *Green* is dyed thus: first the yarn is dyed yellow with Saw-wort (*Serratula*) [*Serratula tinctoria* L.], and then blue with *Indigo,* since yellow and blue together make green. *Sea-green* was dyed with diluted acetic acid and salt in an untinned copper kettle, that was left till it became verdigrised. Then the yarn was put into it and stirred often to make it evenly dyed, but it was not boiled.

The credit system that the townsfolk were carrying on with the farmers was described to us by a merchant who traded here; the merchant yearly grants the farmer credit and thereupon gives him food and drink and goods when he comes to town, lends him money to pay his taxes, often credits him with more than he can hope to get back. The obligation of the farmer is to take everything he brings to town to his own merchant, without asking for cash; if a farmer, after such trade over several years, wished to change, he would be ruined. The farmer does not argue over his goods with the merchant, who himself sets the price, and should the farmer dare to make a deal with another merchant, he would at once be bankrupted.

We offered to teach the shepherd here which plants are dangerous to sheep, *Flammula, Equisetum, Anthericum, Mercurialis, Juncus, Myosotis,* as well as the good sheep-grass, plants which all good shepherds should know, but he had no time for it.

The plants growing in the meadows around Gothum church were the following: *Herminium monorchis dictum* [*H.* 740 *Monorchis* of Index = *H. monorchis* (L.) R. Br.], *Ophioglossum* [*O. vulgatum* L.]. *Tordylium quod caucalis semine aspero flosculis rubentibus* [*T.* 224 *arvense* = *Torilis japonica* (Houtt.) DC.], *Chaerophyllum caule maculato* [*C. temulentum* L.], *Cichorium* [*C. intybus* L.], *Scutellaria, foliis integerrimis* [*S. hastifolia* L.], *Agrostemma* [*A. githago* L.], *Sphondylium* [*Heracleum sphondylium* L.], *Agrimonia* [*A. eupatoria* L.], *Dulcamara* [*Solanum dulcamara* L.], *Thalictrum caule folioso sulcato, panicula multiplici erecta* [*Th.* 453 *palustre* L. of Index = *Th. flavum* L.], *Thalictrum caule folioso equali, foliolis caulinis acutis panicula simplici, floribus nutantibus* [*Th.* 454 *arvense* of Index = *Th. kochii* Fries], *Hydrocotyle*

[*H. vulgaris* L.], *Orchis bulbis indivisis nectarii labio quinquefido punctis scabro* (*It. Oel. 45*) [entry of June 2; *O. militaris* L.], and *Orchis nectarii labio quadrifido punctis scabro* (*It. Oel. 45*) [entry of June 2; *O. ustulata* L.]; both these orchids grew all over the place together with *Aster montanus lutea, salicis glabro folio C.B.* [*A.* 696 *salicifolius* of index = *Inula salicina* L.].

Serapias quae Helleborine latifolia montana C.B. (June 28) [*S.* 734 *Helleborine* of Index = *Epipactis helleborine* (L.) Crantz] was in flower now with a straight spike: *Germen obverse subulatum: Petala 3 exteriora striata, rudiora, lanceolata: 2 vero interiora conniventia, ovata, acuminata, subpurpurea s. incarnata. Labium inferius trifidum; laciniis lateralibus erectis, acutiusculis, rubro striatis; intermedia vero subcordata, integerrima alba: macula ad basin cordata, elevata, tricuspidi. Nectarium obsoletum, convexum, seu ventricosum, minime gibbum.*

We had to spend the night at Gothum, since neither we nor the sheriff succeeded in obtaining horses, although we had asked for them at noon.

July 3 [p. 225]

We now left Gothum for Östergarn.

We had coniferous forest, mainly of firs, without any deciduous trees, all the way between Gothum and Östergarn. In some places there had been accidental fires 20-30 years ago, which could be concluded from the small amount of fully grown pines; one could also see what happens to the soil when it is burn-beaten: the earth and the limestone bedrock were covered with a small amount of limestone gravel and sand, the earth was still bare and nothing more than *Helianthemum* [*H. chamaecistus* Miller], *Sedum floribus albis* [*S. album* L.] and *Sedum acre* grew here.

Sedge fen, with extensive swamps, lay on both sides of the road.

Euphorbia or *Tithymalus Oelandicus* (*It. Oel. 91*) [entry of June 8; *Euphorbia palustris* L.] grew in a meadow to our right, ¾ of a mile from Gothum.

Scirpus culmo striato, spica bivalvi terminatrice longitudine calicis, radicibus squamula interstinctis or *Scirpus folio culmi unico* [*S.* 42 *cespitosus* of Index = *S. cespitosus* L.; apparently a mistake in identification by Linnaeus for *Blysmus compressus* (L.) Panzer ex Link, syn. *S. compressus* (L.) Pers. non Moench] also grew here. This alone forms almost all the tussocks in the marshes.

Anga church, in the parish of Kräklingebo, stood on our left from the main road.

Verbascum, also called *Tapsus barbatus* [*Verbascum* 186 *Tapsus* of Index = *V. thapsus* L.] grew abundantly in this area. Pharmacists should use this plant for *flores, aquam & oleum Verbasci,* but another plant for *radicem Verbasci.*

At Kräklingebo church 1¾ miles from Slite we took a rest and visited the church and the meadows.

A snake [*Natrix natrix* (L.)], which was caught here, was black with white spots on the sides of the belly, the spots being on the sides of scales which covered the abdomen; the chin was white, at the ears yellow; temples white with black transverse lines; scales on the back had a upraised line; the whole underside gleamed in the sun; teeth were small; thus it belonged to the grass-snake genus.

Torsborg, the only mountain in the district, was situated more than ⅛ of a mile from Kräklingebo; it was a big, high, steep, bare and harsh mountain flat on the top. On the north and west sides it was precipitous like a wall, and at the bottom, which was broad, high and sloping, there were caves, often to 26 ells [52 feet] deep. When one stood on the top of this high Torsborg, the tallest forest looked like a flat field, and one could easily count 30 church towers. The open space on top of the mountain was almost horizontal, some 2000 ells [4000 feet] long and equally broad; on the east and north side the mountain was so steep that no creature except a man could climb it,and then laboriously and only at a few places; on the west side the mountain was for the most part as steep as on the north, but on the south side the slope ended in a field; on this side the mountain was surrounded by a wall, when the steepness did not prevent it. There was a good water supply, which never dries up. If this were a fortification and here and there some loose rocks were blasted away, I cannot see how it could be captured. On the top of the mountain, which had few trees (since it had been devastated by accidental fire long ago) grew *Anthyllis* [*A. vulneraria* L.], often with white flowers; *Trichomanes* [*Asplenium trichomanes* L.]; *Sedum floribus albis* [*S. album* L.]; *Sedum acre*; *Rubia cynanchica* [*Galium triandrum* Hylander]; *Thymus acinos dictus* [*Acinos arvensis* (Lam.) Dandy, syn. *Satureja acinos* (L.) Scheele]; *Serpyllum* [*Thymus serpyllum* L.]; *Helianthemum* [*H. chamaecistus* Miller]; *Asclepias* [*Vincetoxicum hirundinaria* Med., syn. *Cynanchum vincetoxicum* (L.) Pers.]; *Behen album* [*Silene vulgaris* (Moench) Garcke]; *Quinquefolium minus repens alpinum aureum C.B.* [*Potentilla verna* L., i.e. *P. crantzii* (Crantz) G. Beck or *P. tabernaemontani* Asch.]; *Quinquefolium majus repens C.B.* [*Potentilla reptans* L.]; *Erigerum vulgare Fl. Lap.* [*Erigeron acre* L.]; *Spergula foliis oppositis pedunculis simplicibus* [*S.* 378 *nodosa* of Index = *Sagina nodosa* (L.) Fenzl], here and in all the surrounding meadows, more than in any other place. *Galium luteum C.B.* [*G.* 116 *luteum* of Index = *G. vernum* L.]; *Galium foliis quaternis lanceolatis trinerviis, caule erecto* [*G.* 118 *album* of Index = *G. boreale* L.]; *Turritis foliis omnibus dentatis hispidis, caulinis amplexicaulibus* [*T.* 544 *glastifolia* of Index = *T. glabra* L. = *Arabis hirsuta* (L.) Scop.] with quite hairless leaves [var. *glaberrima* Wahlenb.]; *Pilosella major repens hirsuta* [*Hieracium pilosella* L.]. *Ruta muraria* [*Trichomanes ruta-muraria* L.] in all crevices so abundantly that it could supply all the pharmacies in Sweden; *Brunella caerulea magno flore C.B.* [*Prunella grandiflora* (L.) Scholler], was also common here.

Scabiosa, corollatis quinquefidis foliis radicalibus ovatis etc. (July 1) [*S. columbaria* L.] was in flower, from which one could judge that it was only a variant of *Scabiosa pratensis hirsuta,* although otherwise unlike this in leaf and stem. *Origanum* [*O. vulgare* L.]; *Globularia* [*G. vulgaris* L.], in such great abundance that, if needed in Sweden, enough could be found here, was not rare, neither here nor in the surrounding district; *Veronica spicata minor C.B.* [*V. spicata* L.] grew here with narrow leaves. All these which grew here could be sown on the most sterile, dry mountain; for no other place could be drier than this.

Anthericum foliis planis, corollis deciduis (June 26) [*A.* 267 *ramosum* of

Index = *A. ramosum* L.] was in full flower on Torsborgen: *Corolla alba, rotata; petalis lanceolatis; Filamenta sex, alba, erecta, alterna paulo breviora. Antheris verticalibus, luteis; Germen ovatum, triquetrum, poris melliferis singuli angulo germinis innatis, ut in Hyacintho; Stylus albus, erectus.*

Geranium pedunculis bifloris, calycibus pyramidatis angulatis rugosis, foliis quinquelobis rotundis [*G.* 574 *lucidum* of index = *G. lucidum* L.], a beautiful herb, which *Thalius in Hercyn. t. 5* and *Columin in Ecphr. t. 137* well portrayed, but not hitherto found in Sweden, grew especially on the northern side among fallen rocks in the shade of Torsborgen. The root dies after a year; the stalk (and leaves, especially the underside), red, smooth and glossy. The leaves are kidney-shaped, glossy, provided with a stalk, five-lobed, and each lobe three-lobed. The flower-stalk has two flowers. The calyx is somewhat swollen, glossy, with five leaves but three-angled, and each angle marked with three raised lines. The petals are entire and flesh-coloured [see p. 7-9].

Campanula alpina linifolia coerulea C.B., Fl. Lapp. 84 was also found on Torsborgen; its kidney-shaped radical leaves show that it is a variety of *Campanula minor rotundifolia vulgaris* [*C. rotundifolia* L.].

Coronilla corollarum unguibus calyce duplo longioribus [*C.* 690 *Emerus* of Index = *C. emerus* L.], called *Emerus* by Tournefort and hitherto seen growing wild only at Vienna, Geneva and Montpellier, and usually grown in orangeries both in Sweden and abroad, this shrub, which produces so many beautiful flowers, and must not be frosted, was growing wild here, exposed to the cold on the north side of Torsborgen in several places. I would never have believed that this shrub would grow wild in Sweden, even if 20 botanists had said so, had I not seen it myself. *Frutex hic bipedalis: ramis angulatis, viridibus. Folia alterna, pinnata cum impari, tribus quatuorve conjugationibus composita: Foliolis obverse ovatis, integerrimis, subtus pallidioribus. Pedunculi petiolis longiores, laterales, biflori, rariusque triflori. Calyces quinquedentati, bilabiti. Corollae luteae, papilionaceae; unguibus calyce triplo longioribus. Stamina diadelpha.*

Serapias flore rubro [*Epipactis atrorubens* (Hoffm.) Schultes], which grew here, seemed fairly distinct from the one we described at Gothum (July 2), since all five petals were alike in size and colour, and the lower lip formed a sphere somewhat drawn together at the lobes and marked with a raised heart-like spot, but had no lobes at the sides, which upheld the upper lobes.

Papilio hexapus, alis erectis rotundatis integerrimis albis: inferioribus ocellis quatuor superne, septem inferne or *Papilio alpinus Petiv.: Gaz. 37. T. 23, f. 8* [*P. apollo* L. = *Parnassius apollo* (L.)], a large and beautiful butterfly, which is not common in Sweden and very rare abroad, had settled in great numbers in the field on top of Torsborgen, as if very tired, and could not fly away; I do not know if the cloudy weather, the wind or the cold had injured them. The tail had 4 sharp hard claws, almost like the claws of a cat, spreading out a little with a pointed style between them.

Wild horses are caught here on Torsborgen, where one could see trees felled for a kind of fence on the top of Torsborgen as well as on the slopes; two such fences go together like the arms of a seine net, the horses are driven into the narrow part of it and are thus trapped.

We went from Torsborgen to Östergarn.

One had the opportunity here of seeing how sedge-roofs were thatched.

From the ridge to the eaves were laid slender pine trunks with branches which, on the side turned upwards, are cut off a quarter-ell [6 inches] from the stem. These trunks lie a quarter-ell apart from each other and are supported by the horizontal roof beams. The sedge is hung thickly on the cut-off branches, and not fastened in any other way, since the trunks are laid with their base on the roof edge, the branches thus keeping the sedge in place. A sedge roof would last 20 years if the sparrows did not destroy it in their search for grubs. The year it has been cut, the sedge does not put forth a single leaf, but stands there with withered stubs; the longer it is left, the better it grows, and it cannot be harvested more often than every fifth or seventh year. It does not bear fruit if it is not left untouched for a long time, for which reason we searched in vain for its spikes on very many harvested sedge marshes until we at last found it.

Some say that the stones are softer when they lie in the earth than when they are exposed, others deny this. It is well known that the chalk stones in Flanders, which are used as a building material, become harder in time than when they were cut in the quarries. The quarriers here told us unanimously that the sunstone, or limestone which lies on the surface, is harder than the limestone taken some ells lower in the quarry, for which reason the quarriers often transport their limestone by long and laborious roads, since they cannot use the sunstone, although it may lie in the vicinity of their lime kiln.

We spent the night at Östergarn.

July 4 [p. 231]

The church of Östergarn is situated on the east shore, in the centre of Gotland, where the island is broadest; from here to Västergarn the land is 5 miles wide and from the north cape of Gotland to the south cape it is 15 miles long.

The mountain near the church was high and steep on all sides, as if it had been a son of Torsborgen. From the top of this mountain there is a delightful view to the north-east; one saw from here the church at the foot of the mountain, then the beautiful fields and meadows, the green woods, other steep smaller mountains, bays of the sea and the blue sea as far as to St. Olofsholm, so that it would hardly be possible to find a more lovely place for a summer house anywhere on this island.

Some big ants [*Formica herculeana* L. = *Camponotus herculeanus* (L.)] were running about here; the body was almost as big as a bee, quite smooth and mostly black except for the abdomen towards the breast and the breast itself, which were yellow-brown. The jaws had five teeth or incisors, with which it could bite fairly hard. The antennae consisted of 13 segments each, of which the basal one was very short, the second as long as half the antenna, the 11 other segments short and of equal size. The scale on the back between the breast and the stomach was oval, entire and not hollowed out. This is thus to be regarded as a separate species and not the female of any ant which builds hillocks.

Sanguisorba spicis ovatis [*S.* 130 *rubra* of Index = *S. officinalis* L.], which has until now only been seen in our finest gardens, grew here in the meadows between the sea and the church, here as well as near the church of Alskog; we had never expected to find it growing wild in Sweden. The spikes were rounded or oval, bright red on the petals, stamens and pistils; each leaf consisted of 3-5 pairs of leaflets with a solitary leaflet at the top; the leaflets were heart-shaped and deeply serrated.

The plants in the meadow were *Ophrys foliis cordatis* [*Listera cordata* (L.) R. Br.]; *Lathyrus, qui clymenum parisiense* [*L. palustris* L.]; *Convallaria foliis cordatis* [*Maianthemum bifolium* (L.) Schmidt]; *Aira quae gramen lanatum Dalecampii* [*Holcus lanatus* L.], a rare grass in Sweden; *Scutellaria foliis integerrimis* [*S. hastifolia* L.]; *Chelidonium* [*Ch. majus* L.], *Ranunculus chelidonicum minus dictus* [*R. ficaria* L.]; *Alliaria* [*A. petiolata* (Bieb.) Cavara & Grande]; *Papaver rhoeas*; *Plantago foliis semicylindraceis integerrimis* [*P.* 127 *anguina* of Index = *P. maritima* L.]; *Hydrocotyle* [*H. vulgaris* L.]; *Carex spicis linearibus erectis, mascula breviore inferioreque: bracteis aphyllis, capsulis distantibus* [*C.* 758 *spica varia* of Index = *C. digitata* L.]; *Stachys sylvatica Riv.* [*S. sylvatica* L.], all in flower; *Anthyllis flore rubro* [*A. vulneraria* L. var.] was growing on the rocks.

Ficaria or *Chelidonium minus* [*Ranunculus ficaria* L.] which was growing abundantly in many places, seemed to drive away all other herbs, including ramsons (June 25).

The fields were coloured purple by *Melampyrum arvense Riv.* [*M. arvense* L.], and *Rhinanthus* [*Rh. crista-galli* L.], growing more abundantly here than in any other place, made the ground look all yellow.

We found many kinds of corals by the sea shore such as *Milleporae* and *Madreporae simplices*, but less *Madreporae aggregatae* than at Kappelshamn.

Petrifactions were mixed with the corals on the shore in enormous quantities, such as *Entrochi, Strombitae* which were like screws, *Cornua-ammonis*, smooth on the inside, *Conchitae laves caput sepentis referentis*, etc.

The farm houses were remarkable in that they had one window in the gable and one in the wall, but the window was not placed in the middle of the gable but closer to the corner; this gave a very good light on the table that was placed in the corner between the windows, while the bed in the diagonal corner was very dark and free from draught. The walls were made of sawn boards, three boards from one tree trunk, which were not joined to each other at the corners, but placed in upright cornerposts as in a barn.

Stalactites were found here and there in the crevices of the mountain, though not very big.

Gammelgarn church was passed on the right hand on the way from Östergarn to Alskog.

Schoenus culmo tereti nudo, capitulo ovato, involucri diphylli valvula altera subulata longa or *Schoenus flosculis spicatis* [*S.* 36 *capit. major* of Index = *S. nigricans* L.]. This grass, not discovered before in Sweden, was found in the sedge marshes. *Spica ferrugineo-nigricans involucro s. gluma universali bivalvi; valvulis emarginatis cum acumine intermedio; horum acumen valvulae exterioris semicylindraceus, viridis apice nigro, subulatus, Spica triplo longior; valvulae minoris mucro Spica nullo modo longior. Spicae plures, fasciculatae, lanceolatae, nigro ferruginae in unicam spicam ovatam collectae; ab involucro universali includebantur. Singula spiculae constabant calyce biglumi & flosculis paucis triandris monogynis, stigmate trifido, non vero bifido* [see p. 12].

A meadow with unusually tall grass lay below a farm; when we investigated the cause of this, we found that the dung hill was so placed in the yard that the rain water carried the dung into the meadow.

The farm houses with white chimneys and surrounded by tall trees showed us how greatly trees planted around the farms adorn them.

Tall heather, blueberries and other plants as in Sweden grew in the forest.

A bush with small ell-long [2 feet] branches grew by the road; it looked rather like a *Cotoneaster folio rotundo non serrato C.B.* [*C. integerrimus* Medicus], and maybe it was one, but the leaves were bigger and more downy on the underside, and it had neither flower nor fruit, so we had to leave it in doubt.

Alskog church is built in a place where nature has made a landmark between the north and south districts. All the northern part of Gotland that we had traversed was limestone bedrock and a hard and dry soil with pine forest; but here where a new land began, which extended southward, the soil was sand or clay and mould without rocks; by the villages and meadows there were more deciduous trees, oak, ash, birch, hazel, hawthorn and blackthorn. This more southern and milder district consists of the following parishes: Alskog, Garde, Lye, Stange, Hemse, Fardum, Leivede, Eistad and others to the south; but the northern part has the parishes: Buttle, Eitelem, Linstad, Garum and Frojel.

The characteristic flowers were *Jacea vulgaris major laciniata* Tournef. [*Centaurea scabiosa* L.], *Orchis muscam referens* [*Ophrys insectifera* L.], *Jasione* [*J. montana* L.], *Cornus foemina* [*Cornus sanguinea* L. = *Swida sanguinea* (L.) Opiz] and *Gramen spica brizae majus* [*Brachypodium pinnatum* (L.) Beauvois].

Several "fairy rings" were seen in the little fields between the shrubs in the meadows. The fairy rings are greenish-blue rings of *foliis Cynosuri* [leaves of *Sesleria caerulea* (L.) Ard. var. *uliginosa* (Opiz) Hegi], the same as on Öland (*It. Oel. 66*) [entry of 4 June].

We had planned to visit Lye church, in the parish of Alskog, to read a runic stone of which Wormius has made an imperfect drawing in *Monum. Dan. 454. L. 5, c. 5.* But since the Rev. [Lars Nilsson] Neugard in Östergarn had copied the text accurately and given it to me, we did not go to Lye, which would have meant a detour. This is the text of the runic stone in Lye:

ᛁᛆᚴᚢᛒᚱ·ᚽ·ᛚᛁᛐᛚᚱᛆᛚᚢᛘ·ᚼᛆ·ᛚᛁᛐ·ᚴᛁᚱᛆ·ᚦᛁᚿᚿᛆ·
ᛌᛐᛆᛁᚿ·ᚢᚠ·ᛁᚱ·ᚠᛆᚦᛁᚱ·ᛌᛁᚿ·ᛆᛚᚠ·ᛆᚴ·ᛒᚱᛆᚦᚱ·
ᛌᛁᚿᛆ·ᛚᛁᚴᛆᚢᛁᚦ·ᚯ·ᛌᛁᛘᚯ·ᛒᛁᚦᛁᛘ·ᚠᚢᚱᛁ·ᚦᛆᛁᛘ·
ᚯ·ᛆᛚᚢᛘ·ᚴᚱᛁᛌᛐᚢᛘ·ᛌᛁᛆᚢᛘ·ᚯ·ᚦᛆ·ᚢᚱ·ᛚᛁᚦᛁᛐ·
ᛆᚠ·ᚠᚢᛘ·ᛒᛅᚱᚦ·ᚠᛁᛁᚱᛐᛏ·ᚼᚤᚱᚦ·ᚱ·ᚯ·ᛆᛁᚿᚢ·
ᛆᚱᛁ·ᛘᛁᚿᚿᛆ·ᚦᛂᚿ·ᚡ·ᛐᛁᚼᛐ·ᛁᚱᚯᛁᚦᛁ·ᚱᛁ·ᛒᚱᛁᛘᚦᛁ·
ᚡ·ᚯ·ᚱ·ᛌᚤᚢᛐᛆᚼᚱᛁ·

Linnaeus: *iakauper . i . litlaronum . han . lit . giaa . þinna .*
stain . yfr . faþir . sin . olof . ok . brouþr .
sina . liknuiþ . ok . simon . biþim . furi . þaim .
ok . alum . kristum . sialum . ok . þa . uar . liþit .

af . gum . byrþ . fiirtan . huntraþ . or . ok . ainu .
ari . minnna . þen . V . tihi . ir . ok . iþi . ari . brimaþi .
k . ok . r . sunutahr . i .

G 99: *+ iakaupr . i . litlaronum . han . lit . giara þinna . stain . yvir .*
faþur . sin . olaf . ok . broyþr . sina . liknuiþ . ok . simon . biþim .
fi þaim . ok . allum . krisnum . sialum . ok . þa . uar . liþit af .
guz . byrþ . fiurtan . huntraþ . ar . ok . ainu . are . minna . þen . V .
tihi . ar . ok . i . þi . ari . brimaþi . k . ok . r . sunudahr . i . XI .
raþu .

"Jacob in Lilla Rone had this stone made in memory of his father Olav and his brothers Liknvid and Simon. We may pray for them and all Christian souls. And then had passed from the birth of God fourteen hundred years and one year less than fifty years. And in that year the rune *k* was prime number and the rune *r* Sunday letter in the eleventh row."

This stone is unique, in that it has a year-number, namely 1409.* Thus it is made during the reign of Queen Margareta; it can also be seen from the letters, which are very condensed, 2 or 3 often being written in the same character, that this stone is one of the most recent. Litlaronom or Lilla Råne is a farm in the parish of Lye.

We spent this night in Garde.

July 5 [p. 237]

Garde church did not show us anything very remarkable except concerning the soil, which was much richer here, as it is further south, with more and lighter black mould and greener and leafier pastures.

Eitlem [now Etelhem] church, the annex of Garde, was on our right hand ½ a mile further away.

När church was ½ a mile yet further. We spent the Sunday there.

Professor N. Norby†, the pastor here at När, was a thoroughly learned man, immensely well read, such as we had not expected to find here "lying among the bushes"; he had travelled much and endured much, and his learned discourses kept us here.

The farmers' botany is not to be despised, and the farmers, at least here, have their own names for almost all plants. I brought a good-natured farmer with me to the meadows, and he knew far more plants than I would ever have expected, and his names for them had often very nice origins.

Monorchis [*Herminum monorchis* (L.) R. Br.] "Desmansblomma" [musk flower], because of the form of the root and the smell of the flowers.

Dianthus barbatus "Sarons Blomster" [flower of Sharon]. *Anemone nemorosa* "Fagelblomma", from the verb fuga, i.e. collect brushwood from the pastures, since it flowers when this is done.

Hepatica [*Anemone hepatica* L.] "Killinge-Blomma" [chicken flower], it flowers when the chickens come. *Primula lutea* [*P. veris* L.] "Gök-Blomma",

---

* There seems to be a discrepancy here, but 1409 is undoubtedly correct as Queen Margareta died in 1412–*Ed. Sec.*

† This was Nils Olofsson Norby (1675-1742), who had been professor of theology at Stettin (Szczecin), Pomerania, was the author of various learned works and knew Latin, Greek, German and Hebrew.–*W. T. S.*

flowers when the cuckoo calls. *P. Min. purp.* [*Primula farinosa* L.] "Marie-blomma" [Mary's flower]. *Rubus caesius* "Psalmbär" [Psalm berry]. *Opulus* [*Viburnum opulus* L.] "Qvalkebär". *Ulmaria* [*Filipendula ulmaria* (L.) Maxim.] "Brake". *Campanula* [*C. rotundifolia* L.] "Fingerhatt" [thimble]. *Erica* [*Calluna vulgaris* (L.) Hull] "Gränne". *Briza* [*B. media* L.] "Bäfwegräs" [trembling grass]. *Fragaria* [*F. vesca* L.] "Rodbär" [red berry]. *Agrostemma* [*A. githago* L.] "Slätt". *Bromus 1* [*B. secalinus* L.] "Gåshafre" [goose oats].*Behen album* [*Silene vulgaris* (Moench) Garcke] "Tarald". *Ophioglossum* [*O. vulgatum* L.] "Laketunga" [healing tongue]. *Myagrum* [*Camelina sativa* (L.) Crantz] "Will-Lin" [wild flax]. *Melampyrum seget.* [*M. arvense* L.] "Puthwete" [drum wheat]. *Pteris* [*Pteridium aquilinum* (L.) Kuhn] "Fräkna". *Hypericum* [*H. perforatum* L.] "Hirkenpirk" *ab hyperico. Millefolium* [*Achillea millefolium* L.] "Pestilenz-Blomma" [pestilence flower]. *Trifolium* [*T. pratense* L.] "Honungs-Blomster" [honey flower]. *Agrostis spica venti* [*Apera spica-venti* (L.) Beauv.] "Tåtel".

Mustard, the same as we use with our food, grew in the fields; the farmers assured us that the seeds do not become bitter at this place when they ripen.

"Gothlands-Rofwor" (*Brassica, quae Napus sylvestris*) [*B. napus* L., rape] was found in the fields, but surprisingly close to the seashore; and is called "Åker-Rofwor" here. This fine turnip is much better to eat than others, although smaller, provided that it is harvested before it develops a stem.

Pinguicula [*P. vulgaris* L.] is called "Fetnacke" [fat-neck] here; the farmers boil it in water and use the decoction for washing the heads of their children, which kills the lice and makes the hair grow.

"Pänninge-Gräset" (*Rhinanthus*) [yellow rattle] was heard rattling its fruit when we walked through the meadows, which, as in other places, is a sign for the farmers that it is time to mow.

"Madra" (*Rubia cynanchica, It. Oel. 115* [entry of 12 June]) [*Galium triandrum* Hylander] grew here almost exclusively in dry and sterile spots and where there was clay. Wherever we have travelled in Gotland, we have seen this plant in such quantities that no herb has been more common, so if it is to be collected somewhere for dye-works, this is the place to do it. With the roots everyone here dyes wool red; one cooks them with the sourest beer, especially the kind called "Standebilla", for the sourer the beer is, the more intense the colour becomes; after boiling the roots, the yarn or the stockings are put into the decoction while it is still warm, then it is rinsed quickly in a weak lye. Here should be noted, according to the farmers, it is essential that the roots be collected before the cuckoo starts calling, that is before the roots put up stems; since the roots are looser then and yield more colour.

"Gullands-Korn" was the peasants' name for their two-rowed barley or *Hordium disticum* [*Hordeum distichum* L.], that is the most commonly used grain here. When we had left the northern hilly district of Gotland, where rye was chiefly sown, and had come to the southern district, where the soil is looser, here we saw more barley than rye.

My companions amused themselves playing "pärk" in the afternoon, which is a very nice ball game, often played on Gotland by the farmers, who taught us to play it. It is not seen in Sweden, although it is common in Holland and in Gotland; it demands fair agility and speed.

During the night we stayed at När.

July 6 [p. 239]

Today we went from När to Burs.

The land looked very infertile, being of a bare, cloddy and cracked clay, which is waterlogged during spring and autumn; the little grazing available is said to be very good, consisting almost exclusively of sheep's fescue and *Serpyllum* [*Thymus serpyllum* L.]. The grazed fields were white with *Filipendula* [*F. vulgaris* Moench]. Pine grew in the forest, but no fir, and there were more deciduous trees in the pastures than in the north.

The hills, which were so dry that it seemed as if no plant could grow on them, nevertheless had their own herbs, which were not often seen in other places, such as *Herniaria* [*H. glabra* L.], *Serpyllum* [*Thymus serpyllum* L.], *Acinos* [*Satureja acinos* (L.) Scheele = *Acinos arvensis* (Lam.) Dandy], *Lagopus* [*Trifolium arvense* L.], *Gallium luteum* [*Galium verum* L.] and *Cinquefolium argenteum* [*Potentilla argentea* L.].

Herniaria [*H. glabra* L.] was specially remarkable for its flowers in which besides the five stamens there were five more filaments, alternating with the petals, but without anthers.

The plants in the meadows at Burs were above all the musk-scented *Monorchis* [*Ophrys monorchis* L. = *Herminium monorchis* (L.) R. Br.], growing profusely; *Carex spicis ovatis pendulis: masculina longiore erectiore, caule repente* [*C. limosa* L. from the diagnosis, but probably *C. flacca* Schreber]; *Carex spicis pendulis: mascula erecta, foemineis oblongis distichis, capsulis nudis acuminatis* [*C. capillaris* L.]; the seeds of this plant are slightly separated from each other when the bracts have fallen off. *Carex spicis pendulis; mascula erecta, foemineis ovatis imbricatis, capsulis confertis obtusis* [*C. pallescens* L.]. *Carex spica simplici androgyna* [*C. pulicaris* L.], on which the upper part of the spike, where the male flowers had been, was already withered; the stalk was not triangular but rounded and hollowed on one side. *Orchis muscam referens* [*Ophrys insectifera* L.]. *Orchis hiante cucullo major* [*O. militaris* L.].

The cultivation of oil seeds, often practised in Flanders, Brabant, etc. to the great advantage of the inhabitants, has not been tried by us, although there is good reason for it in this place where the fields are overgrown with wild flax, charlock and rape, and the oil pressed out of their seeds would be quite as good as the foreigners' *Kohlsad* und *Rapsad.*

Rape (*Napus sylvestris*) [*Brassica napus* L.] grew abundantly among the barley at Burs, although it was far from the sea; the root was as thick as a carrot or a turnip; the leaves were smooth and clasped the stalk as on kale; the sepals were somewhat outspread; the pods were big, as were the seeds, which would yield much oil if they were pressed. No weed can be better for this, as it cannot easily be eradicated from the fields and thus would certainly be easy to cultivate for oil.

Charlock (*Rapistrum flore luteo C.B.*) [*Sinapis arvensis* L.] was here so abundant that we saw a whole field yellow with it; the barley was so suffocated that it had not reached a quarter-ell [15 cm] in height; had we not been told otherwise, we would have believed that the charlock had been sown here and not the barley. We tried to persuade the farmer to press oil from the seeds, but he pointed out a difficulty, namely that the pods do not all ripen at the same time, but one plant drops its seeds while the other is still preparing them. The root of the charlock is thin, the leaves scraggy, uneven and divided; the sepals are spread out and bent back.

"Wild-linet" (*Myagrum*) [*M. sativum* L. = *Camelina sativa* (L.) Crantz] was growing among the flax and so smothered this that the farmers had to weed it out by much effort; it should thus be easy to cultivate as well; in several places in Germany, etc. whole fields are cultivated for the oil pressed from the seeds.

Burs church was visited at noon.

Bellis officinar. [*B. perennis* L.] was growing everywhere by the roads, and sometimes *Angallis flore rubro* [*A. arvensis* L.].

Pear trees with big spines and sour fruit were growing wild in the forest.

Aira quae gramen pratense, spica lavendulae [*A.* 71 *violacea* of Index = *A. caerulea* L. = *Molinia caerulea* (L.) Moench] had a little style between the two flowers, but there was no stigma as there is in *Melica.*

The "Swedish hay-seed" I have described in the *Handlingar* of the Academy of Science, 1742, p. 191, where I have shown the advantages of *Medica sylvestris floribus croceis J.B.* [*Medicago falcata* L.] over other "grasses" that are used for hay in Sweden; this observation I made for the first time here near Burs church, where between the fields was an elevated, barren and white piece of land on a hill, upon which nothing would grow except *Phleum* [probably *P. pratense* L.] and a few plants of *Valeriana Locusta dicta* [*Valerianella locusta* (L.) Betcke]. On this very dry field there were big green bushes of this *Medicago,* that showed us how the plant can flourish in the poorest soil. The plant was growing on the banks where they were narrowest and where the scythe cannot be used until the crops are harvested; we learned from this and were strengthened in our opinion why the plant is not more widely spread, namely the roots do not survive more than 4-5 years and the seeds ripen rather late in autumn, long after the hay-making season is over; those who want to sow this "grass" in their meadows should thus leave it uncut every third year till the crops are harvested, and the plant has sown its own seed. Lucerne and Sainfoin, of which the seed is imported every year and which is a sister of our "hay-seed", does not endure our winters as well as this and, although it may be incomparable for hay, it has no advantage compared to ours; the foreigner requires rich soil and hot summers, but the native plant endures everything and flourishes on the poorest soil. Thus, as the proverb says, we "cross the fresh stream to fetch rotten and stagnant water" when we buy foreign "hay-seed" and despise our own.

The seeds of the plant are gathered and stored in a not too warm room during the winter and are cast on the ground, with pods and all, without any special regard to the soil (except that it should not be too wet). The seed is rubbed through the moss into the earth with a mattock, a rake or a harrow; it is mowed twice yearly, but every third year only once after it has borne fruit; it will endure our winters, there is no need to buy it; and it needs neither manure nor re-sowing after some years. It is the best possible food for cattle and often reaches a height of 2 ells [4 feet, 120 cm]; the shoots from a single root can yield an armful of hay, even upon the poorest sandy soil. Personally I rate this single discovery so high, that I consider it sufficient to pay all the expenses of my journey to Gotland. Those of my countrymen who want to see this plant could ask anyone studying at Uppsala University, who should be able to show it to him, since it is demonstrated to the students of Uppsala every summer; if the student cannot, it is his fault, not mine.

We visited Stange church, which stands a ½ mile from Burs, of which it is an annex.

Household remedies in this place were *Ophioglossum* [*O. vulgatum* L., adder's tongue], which the farmers use as a dressing for leg ulcers, since it removes inflammation and heals them quickly. The dry flowers of *Behen album* [*Silene vulgaris* (Moench) Garcke] were used by the farmers to put on erysipelas in the same way as doctors use elder blossoms; the same flowers are burnt when restless children cannot fall asleep.

A common feature at the tables in the farmhouses were benches like canapeer [couches] with a reversible support for the back enabling them to be used from either side; they were all very neatly made [see Plate 8].

"Standebilla" is the name for a kind of weak beer that the farmers sometimes brew here; they put mash or weak beer in a large wooden vessel and leave it to ferment, which yields a sour drink, but the fermentation clears the liquid; as soon as some of it has been drunk, fresh water is added to it, and thus it provides sufficient drink for three months; one might believe that this drink would be worse than water, but since fermentation precipitates the solid particles it must be superior where the water contains lime.

Flies [*Musca urticae* L. = *Ceroxys urticae* (L.)] with outspread wings, which had three transverse black stripes on each wing, sat on the nettles.

Evening fell during the return journey to Burs, where we spent the night. The chafers [*Amphimallon* or *Melolontha* species] were buzzing and whirling around our faces, as if they meant to fly into our eyes. They were of the smaller variety with three white stripes on the wing-cases.

The nightjar (*Hirundo caprimulgus*) [*Caprimulgus europaeus* L.] was churring, and *Ortygometra* [*Rallus crex* L. = *Crex crex* (L.), corncrake] was croaking in the fields.

July 7 [p. 245]

Today our journey was from Burs to Råne and Grottlingebo.

We saw several sedge marshes by the road; here we found the flower of sedge [*Schoenus mariscus* L. = *Cladium mariscus* (L.) Pohl] for the first time; it consisted of a *Calyx squamis plurimis imbricatus, biflorus: Flosculis dipetalis, diandris, monogynis, stylo semi trifido.*

There were several burial mounds along the road.

After ½ a mile's journey we came to Råne church.

The inhabitants of Råne are famous for their old-fashioned language and costumes, for they are said never to have imitated the habits of foreigners, but time has changed them too, so that we had difficulties in finding their white knee-long shirts, the black vest that is laced outside the blouse, the baggy trousers and the shoes made to suit both feet.

Darnel (*Lolium verum*) [*L. temulentum* L.] was growing in the barley fields; all agreed here that beer made from barley mixed with darnel makes the guests dizzy and mad and also while drunk they become blind.

"Jord-bi" [earth bees, *Apis rostrata* L. = *Bembex rostrata* (L.)] * were flying here and there along the road as if in swarms; they are wild bees that make nests and holes for their young in the sandy soil, but with not more than one grub in each, as big as an acorn while it is still in this casing. The bees themselves are as big as hornets; eyes, mouth and feet pale yellow; breast and abdomen black; every abdominal ring is marked with a pale yellow transverse wavy line; the first ring has a pale yellow spot on each side, the last one a pale yellow point; the wings are glaucous with many veins. The upper lip of the mouth is large, bent inwards, hollowed below, pointed. The jaws were in a pair, pale yellow, with black points encircling the upper lip; the antennae were black above but yellow below; at its tail was a short, stiff and horn-like sting. There were two sexes of these bees; one sex is as we have described, but the other had a sting like an ordinary bee, which was long, flexible and pointed. This one, however, was spotted white on the breast and the abdomen where the other had yellow markings, and the antennae were all black.

The "ettermyrorna" [*Formica rubra* L. = *Myrmica rubra* (L.)] had just hatched their males, which were not black-brown like the ants themselves but were dark, with a paler colour on the back of the breast behind the wings, and on the feet and the back.

Dragonflies or *Libellulae* of the smallest and copper-coloured kind [*Libellula puella* L. = *Coenagrion puella* (L.)] were caught, hanging together. The eyes, the sides of the breast, the first and last segments of the abdomen and the entire abdomen below in the male were pale grey; he had four claws on the tail with which he grasped the female around her neck. The female was similar, but the eyes, the sides of the breast and the abdomen were flesh-coloured below; under the tail there were two claws, straighter than the male's, with which she grasped him under the breast. Both male and female had small, dark red grain-like projections under the breast.

The meadows were full of *Aira, quae gramen lanatum Dalechamp.* [*Holcus lanatus* L.] ; ramsons [*Allium ursinum* L.] grew luxuriously under hazel and ash

---

* Råne, Gotland, is the type-locality of *Apis rostrata* L., *Syst. Nat.* 10th ed. 7: 577 (1758), *Fauna Suec.* 2nd ed. 422 no. 1700 (1761).–*W.T.S.*

trees. *Cichorium* [*C. intybus* L.] grew abundantly enough for all pharmacies. *Campanula Trachelium vulgare dicta* [*C. trachelium* L.] was common.

Porrum or "käipe" [*Porrum* 266 *Kaipe* of Index = *Allium scorodoprasum* L.] occurred here with its straight stalk and a head of small bulbs, between which grew out small purple-coloured flowers on long flower stalks, with six white stamens, of which three were pointed like a bristle and two-cleft with the anther between the divisions, the alternate stamens being undivided; the style was stout and short [see p. 54].

Forests were fairly thick with pines.

Elder (*Sambucus*) [*S. nigra* L.] was seen in the gardens.

Goats were common here.

The soil was sandy all the way; the meadows were often more delightful than in the north, although in some places they were not well kept and yielded less hay.

Eke church, in the parish of Råne, and ½ a mile from it, was seen on our right; around it were vast meadows and very leafy groves.

The rye was usually splendid, but somewhat thin.

Grottlingebo church lies one mile from Råne and is one of the most outstanding in the whole country. A rune stone, which was easily read, lay on the chancel floor, and another one stood in the chancel wall, looking like a seat.

The vicarage of Grottlingebo like most vicarages in the district was an old stone house with heavy inner and outer walls of stone filled between with sand and rubble and, when a stone was taken away, the sand poured out from the wall.

A household remedy for swollen and oedematous legs was marsh moss or *Sphagnum palustre molle* cooked with beer and laid on as a dressing.

We went out to Grottlingebo point and tried to reach two little islets beyond it, but the north-west wind blew too strongly. From this cape we saw several churches, most of them in the south, since this parish borders on a plain that ends at Hoburgen.

Sheep graze here in the open both winter and summer without ever being taken under a roof; nevertheless they were very big. Most of the rams have two horns, but some have four and even six, of which the lateral horns were bent, the middle horns straight; for the fibres in the horns pull to one side, so that the divided horns bend towards the thicker side while the inner and weaker side yields.

Juniper bushes here were many and low-growing; their fruits were said to ripen only every sixth or seventh year, but the unripe berries gave as strong gin as the ripe ones.

We saw several quarries on this cape; they were usually covered to a height of two quarter-ells [1 ft, 30 cm] with sand and *Conchis albis striatis* [white striped shells]. They consisted of stratified limestone to a depth of 2 ells [4 feet, 120 cm], looking like *Oolithus* or "runsten", with rounded particles consisting of concentric layers of stone. Then followed a dark grey, compact, fine and dry clay, like a *Lithomarga*, which both split into flakes like a schist and broke apart into cubes like a spar; between the flakes were *lamellae* sprinkled with an *impalpable mica* [i.e. "of very fine powder, in which no grit is perceptible when it is rubbed between the fingers".–O.E.D.]; sometimes this clay was quite hard and like fine whetstone. Under this there was a light grey sandstone, quarried for grindstones, which when broken appears wet all through, though it did not lie under water level; after drying it became much harder and denser; one saw clearly that it was derived from sand from the sea; granules, colour and fossil shells confirmed this. When the sandstone is quarried, earth and limestone are removed down to the first layer of clay, then the sandstone is split horizontally with wedges, thus breaking it as far as the perpendicular cracks. The first layer was often 3 ells [6 feet, 360 cm] deep; then there was a separating layer of clay similar to the first, then another layer of sandstone, and so on. With such thin flakes of sandstone the people here cover their roofs, like shingles or slates; the outer corner of it is always turned downwards; this stone is waterproof and lasts for a long time. The sand-mine is not raised above ground but is even and level with the soil; it begins at Grotlingebo and continues out to the south cape.

Garnshamn, an advantageous and beautiful harbour, was situated north of the quarry.

We had often asked the farmers the cause of tussocks, but never received any adequate answer; here, however, we think that we have found an apparent cause: tussocks abound where the land is low-lying and flooded during the winter, while there are none at lower levels which remain waterlogged in spring to the middle of May.

Angantyr's mound (as some people call it; correctly or not is the same to me) was here where the promontory begins in front of Garnshamn; it was made of round stones so big that they could hardly be lifted by one man, and was very high and steep, the circumference being about 113 paces or ells; a ring of stones was arranged around it, about 1½ fathoms [9 feet, 2.7 m] from the mound. Another slightly smaller mound lay in the forest 4-5 gunshots [900-1125 m] northward from the first one. A third one, smaller still, lay one gunshot [225 m] further north. Various small other sepulchral places were seen in the neighbourhood.

Malva caule erecto, foliis subpalmatis obsolete serratis [*M.* 581 *major* of Index = *M. sylvestris* L.] grew in the meadows; we have not seen it before in Sweden, except in Skåne.

Artemisia foliis compositus multifidis tomentosis, ramis floriferis nutantibus (*It. Oel. 112*) [entry of 11 June = *A.* 671 *Seriphium* of Index = *A. maritima* L.] grew on the shore completely white and gave out an incomparably fine scent.

Centaurium minus [probably *C. erythraea* Rafn] was quite common.

Zannichellia [*Z.* 745 *Aponogeton* of Index = *Z. palustris* L.] was found around the quarry. It resembled *Potamogeton gramineo folio* [i.e. *P. gramineus* L.], *sed flores nudi, stamine unico; filamento longo a gemine remoto, Anthera erecta; Pistillo ovato: germinibus 4 conniventibus, stigmatibus extrorsum dilatatis; Pericarpia reflexa, curva, laevia.*

Hirudo depressa alba lateribus acutis (June 26) [*H. octoculata* L. = *Nephelis octoculata* (L.)] was found in a small stickleback or *Gasterosteus tribus in dorso aculeis* [*G. aculeatus* L.], for when the fish was squeezed, out came a big white worm and one could not help wondering that there was room for it in such a small stomach: one might have believed that it were the spawn of the fish, if it had not moved. It was flat, broader at the head, more rounded at the tail, and had neither antennae nor feet; the mouth was hollowed out; many fine transverse lines encircled its body, which were further apart from one another at the hind part.

This night we stayed at Grottlingebo.

July 8 [p. 251]

Today our journey took us from Grottlingebo to Vamlingebo, through Fide, Öja and Hambre.

Large white spiders' webs covered the meadows. One side of the web was woven by the spider like a well or a cylindrical cavern on the ground, where he himself could reside and not be seen by birds or languish in the heat; as soon as his net was touched, he came up to look for his prey. The spider was an *Aranea abdomine fusco ovato, linea ex albida pinnata, cauda bifurca* [*A. labyrinthica* L. = *Agelena labyrinthica* (L.)] ; the breast was a pale grey with three long pale stripes. The belly was dark and glossy with a small white margin, which looked like prickles on the sides; under this oval belly were three pale longitudinal stripes. The tail had two long teeth above and two short ones below.

Vigan was the farmers' name for a district, the boundaries of which they were still quarrelling about, since it had never belonged to any parish. We left the road by the right-hand side, ⅛ of a mile from Grottlingebo, where we saw not far from the road a round hill like a burial mound, in which was a vault, as in a cellar, built of hewn sandstone; the vault was 3 fathoms [18 feet, 5.3 m] broad and long and at least 2 fathoms [12 feet, 3.6 m] high; there was a lot of earth on the floor; the exit was on the south side; in the walls were hewn-out four-cornered holes, like shelves. Earth a fathom [6 feet, 1.8 m] thick lay on top of the vault, mixed with stone and gravel and covered with limestone. This was situated away from the road in a meadow in the wood and was also called "Vigan". Whether it had been a "Castell" [castle] for soldiers to drive away enemies, an "Arrest" [jail] for rascals, a hiding place in times of war, or a granary, I do not know.

Gullbacken was the name of an artificially made hill on the right hand, a ¼ mile from Grottlingebo.

The view from the main road here was somewhat remarkable. To the right the western sea stretched into the land its long Bursvik [Burswik inlet] which ⅛ of a mile from its inner end was cut across by a long sand bank, so that one could ride over it when the water was low, although one must be careful, since the inner side of the bank is rather steep and muddy. To the left the eastern sea was seen and in the south-west the extreme end of Hoburgen.

Fide church, in the parish of Grottlingebo, lay on our left ¾ mile from Grottlingebo.

The ground, which until now had been flat and somewhat convex without high rocks or forest, from here on was more even and green with small tussocks; all the junipers disappeared but the meadows were richer in deciduous trees.

Öja church, a ¼ mile from Fide, had a runic stone that could be read in parts, but on the graveyard there were five stones, of which we could only read one.

The meadows we saw today looked more like groves or gardens than anything else. The trees were oak, more ash, still more birch, but most of all hazel. Where the hazels grew moderately thick, one saw beautiful grass, from which one may conclude that hazel by no means burns the grass. But where the hazel bushes were widely spaced the grass between them was shorter, since there was no shade from the hazels.

The flowers that now graced the meadows were above all *Parnassia* [*P. palustris* L.], *Filipendula* [*F. vulgaris* Moench], *Anthyllis* [*A. vulneraria* L.], *Tormentilla* [*Potentilla erecta* (L.) Räusch.], *Brunella* [*Prunella vulgaris* L.], *Convolvulus* [*C. arvensis* L.], *Cichorium* [*C. intybus* L.], *Myagrum* [*Camelina sativa* (L.) Crantz], *Marrubium* [*M. vulgare* L.], *Agrimonia* [*A. eupatoria* L.], *Tithymalus helioscopus* [*Euphorbia helioscopia* L.], *Ranunculus echinatus* [*R. arvensis* L.], *Hypericum caule ancipiti* [*H. perforatum* L.], *Linum catharticum, Lotus corniculata* [*L. corniculatus* L.], *Alsine gramineo folio minor* [*Stellaria graminea* L.], *Gramen bufonium* [*Juncus bufonius* L.], *Aira spica lavendulae* [*Molinia caerulea* (L.) Moench], *Monorchis* [*Herminium monorchis* (L.) R. Br.], *Helleborine purpurea* [*Epipactis atrorubens* (Hoffm.) Schult.] with *Sphondylium* [*Heracleum sphondylium* L.] in abundance in the uncultivated fields.

Scabiosa: Morsus diaboli [*Scabiosa succisa* L. = *Succisa pratensis* Moench] or scabious, the first autumn flower, raised its blue head, reminding the farmer that the time for mowing had come.

The grasshoppers were chirping in the meadows. We caught one of them, which is called "vårbit" [wart-biter] or *Gryllus cauda ensifera recta, corpore subviridi* [*Gryllus verrucivorus* L. = *Decticus verrucivorus* (L.)]: it is one of the biggest in Sweden; the female draws out her tail like a long sword. The male is entirely green and has four teeth in his tail and two pairs of short claws between the thighs. The wings, which rest on top of one another, are very remarkable in that towards the breast there are two round holes almost the size of a pea covered with a thin transparent membrane. When the grasshopper wants to play and make music for his beloved, he rubs his wings against each other, and the taut membrane makes the sound; thus his voice comes not from his mouth, but from his wings. The female, having no such instrument in her wings, must accordingly remain silent, but she draws a sabre at her tail, which is two-cleft. The mouths of both male and female consist of two pairs of jaws, of which the upper pair has many sharp teeth, but the lower pair are pointed and have no teeth. The very thick tongue is joined to the lower lip, and the upper lip is rounded. When the farmers have warts on their hands, they take such a grasshopper and put the wart to its mouth; the grasshopper bites the wart and spits a black corrosive liquid into it, which makes the wart disappear.

Hambre church, the annex of Öja, was situated ½ a mile further eastwards towards the sea. On the churchyard we read five runic stones. We do not know how it comes about that the names of the people for whom the stones were erected are almost always destroyed.

A gravestone was seen in the churchyard of Hambre with a hewn Swedish text, clear and legible, which told that Master Berren Classon, coppersmith, who died in 1691 was 305 years, but the stone-cutter seems to have counted years as copper-coins.

All day we saw stone walls close to farms as well as to the churches and in the countryside, which are the remains of big houses and vaulted rooms, often 3 storeys high, with small stairs and galleries in the walls. Whether they were built as fortifications or castles for sea robbery, or were houses for monks in ancient times or residences for lords in the Danish time, we do not know; most of the walls have been destroyed, although the buildings would still be fit for use, if the farmers would roof them.

The bedclothes, in which we slept on most of the east side, even in the hottest summer weather, were of beautiful, puffed-out and costly eiderdown.

Our night-lodgings we took in Vamlingebo, ½ a mile from Hamre, for we had journeyed all day in sunshine and in a storm from the north-east.

July 9 [p. 255]

Vamlingebo is the most southerly parish on Gotland. There were eight runic stones in the churchyard at Vamlingebo, but some of them were fragmented and only three could be read.

Today we travelled from Vamlingebo along the eastern shore, which we followed to Hoburgen, and from there back through Sundre to Vamlingebo, and the whole day's journey was sand and gravel, as on a beach. The land was bare and barren for the last ¾ of a mile.

Storms had broken the eastern shore and here and there made pits and holes in the turf like marl-pits, of which the sides were quite steep and sand from them was cast up on to the land. *Centaurium minus* [*C. erythraea* Rafn] grew abundantly on the strand. At the place where we went to sea to cross to Heligholmen, there were chambers made by the sea in the limestone, but without roofs.

Heligholmen is a little island, 504 paces in the east-west direction but 375 paces the north-south, flat, bare and barren, at a distance of 3-4 gunshots [675-900 m] from the mainland. On the southern shore of this island there were rocks 2-3 fathoms [12-18 feet, 3.6-5.4 m] high hollowed out into various *concamerationer* [vaults, cavities] with halls and chambers in them, needing only roofs to make houses of them; nevertheless no-one can dwell here, particularly when the wind blows from the south, except sea-gods and nymphs. *Cochlearia* [*C. danica* L.] grew in the crevices. Eggs of the velvet scoter [*Melanitta fusca* (L.)] were found on the shore. The seagulls screeched in the air and the turnstones (July 2) flew above our heads. A sandbank passes from the island to the mainland, but it was too deep to ride over.

From Heligholmen we followed the eastern shore down to the southernmost cape, where we saw seals sleeping in the sunshine on their stones and great numbers of mallards flying towards the sea. Some pines, at first seen towards the eastern edge, were prostrate and unable to grow straight because of the wind; they all bent their crowns inwards, towards the land. The junipers were little more than a quarter-ell [15 cm] high, often a fathom [1.8 m] broad, quite thick, and looking as if they had been clipped.

The land was hilly and bare, but *Asclepias* [*Vincetoxicum hirundinaria* Medicus] grew in such abundance with its white flowers that it looked as if it was cultivated over the whole area. No animal eats it; if any use could be found for this plant, one could get no better place than here to collect it.

Sheep's fescue and some other herbs growing here were quite small; there was no black mould. We were surprised that the sheep grazing on this meagre soil were fatter than in any other place on Gotland.

The yearly increase of the land was so obvious here, that we could hardly see a better example anywhere else, especially at the eastern shore before the beach narrows and one arrives at the farm. The land, here gently convex, looked on the east side like a ploughed field with furrows parallel to the beach. Each furrow was 1-3 fathoms [1.8-5.4 m] broad, and the side facing the sea was always the broadest. At the beach we could see how the ridges are formed, one for each year, from the gravel thrown up by the sea. The ridges were very distinct close to the sea but the more one went landwards the more difficult they became to discern further up. We walked from the beach towards the land in order to be able to count the ridges without omitting any, and we counted 77 ridges, the last of them at least 500 ells [300 m] away from the sea, as measured with our steps; if we had had instruments, so that we could have compared the 77th ridge with the sea-level and see how much higher it was, we could have ascertained how much the sea has receded in 77 years; I have never seen such an annual recession of the sea, although I have travelled far and wide. It seems strange to me, though, that the 77th ridge is so high above the sea that one can see over the land to the western side; this implies that the cape must be of much more recent origin than one would otherwise believe. It is also difficult to doubt that the sea, which has given such sure landmarks in these ridges, should not make one for every year, especially as they are nowhere discontinuous or joined. It would be desirable to send a land surveyor here to study this carefully.

Farms with fields and meadows lay along the east coast towards Hoburgen. In the meadows grew *Malva caule erecto, foliis subpalmatis obsolete serratis* [*M.* 582 *Alcea* of Index = *M. alcea* L.] and *Trifolium spicis ovalibus imbricatis, vexillis deflexis persistentibus, caule procumbente* [*T.* 618 *sublupulinum* of Index = *T. procumbens* L.], a herb that has not been seen growing in Sweden before. There was another plant in the fields, which I have only seen in Småland before, called *Raphanus siliquis teretibus articulatis unilocularibus* [*R.* 568 *segetum* of Index = *R. raphanistrum* L.]. A fine tree stood here, all alone, braving the storms; it was an apple.

The thrushes (*Turdipilulares*) [*Turdus pilaris* L., fieldfare] flew between the shrubs, and chattered like jays; one of them was shot; it was as big as a blackbird [*Turdus merula* L.]; the head was iron-grey on the upper side and iron-grey likewise the rump above the tail; the feathers covering the back and wings were dark rust-coloured; the feathers in the tail and wings were blackish; the breast and the lower part of the throat were almost rust-coloured and besprinkled with black spots, of which those near the bill were small but more elongated, the lowermost on the other hand broader; feet and thighs were white with a few dark spots; the upper jaw was slightly bent upwards, hardly longer than the lower, dark, but paler in the middle and hollowed out on both sides of the tip; the lower jaw was yellow; nostrils were oval; tongue cartilaginous, sagittate, with an entire margin in the front but as if toothed at the back.

Kliva is the name of a place where the land slopes precipitously towards the eastern sea, with steep cliffs, and in their crevices grew white roses.

Hoburg, a mountain similar to Torsburgen, stands at the extreme end of the land, on the west side of the southernmost point, which goes a little further in the south-east beyond the mountain. This Hoburg is one of the most remarkable things that Nature has made on all Gotland and is very high, like a most beautiful castle, perpendicular on all sides, except that it is in some places narrower at the base than at the top; one cannot climb this mountain except on one side by a narrow pathway. The area on top of the Hoburg is 397 paces in the north-south direction but 145 paces in the east-west, quite bare and flat or sloping slightly downwards towards the middle, so that if one wanted to make a water reservoir, all the rain falling on this area would collect at the middle. The mountain consists of coarse and hard limestone; if anyone would spend a few barrels of gunpowder to blast off some rocks at the path, it would be impossible to come up other than by a rope; thus it could be made into a safe shelter for the inhabitants during times of war, especially in this woodless country. It would also be easy to hew out rooms and chambers in the mountain since it consists of limestone, so that the people could live in them, as in underground cellars, on this area.

We marvelled that so remarkable a place, where Nature, who never makes anything without a reason, has created such a great masterpiece, should not be used for anything, especially on account of its special position at the extreme south cape of the island; moreover there was not even a beacon here, although so many ships are wrecked on these shores, bewildered by storm, darkness and currents, that always mislead seafarers; a beacon could easily be installed in a country which has tar and seal lard in superabundance. In the cliffs were some places full of petrifactions. The west side of the Hoburg was concave and one saw many caves there. On the north side of this great Hoburg, on the western shore, lay a mount like a small Hoburg with caves on its western side. Farther north lay yet another small mount, which had on its western side a very big cave; the room looked like a vaulted chamber extending inwards and quite dark; we did not dare to go far into it because of falling stones. This cave is called the bedroom of the Old Man of Hoburg. The fairy tale about the Hoburg Old Man is known to all children.

Phalaena seticornis spirilinguis nasuta nigra, lineis argenteis transversis quatuor [*Phalaena wilkella* L.* = *Eulamprotes wilkella* (L.), syn. *Argyritis pictella* Zeller] or small moths were in such great abundance on the top of the Hoburg that they threw themselves up like dust before our feet or like the *cicadae* when one walks over water-meadows where the hay has been just mown. They were among the smallest of moths, oblong and black, with four silvery transverse lines, of which the first was bent forwards, the second straight and the third and fourth bent towards the sides; the wings were downy at the edge; the head and antennae were white.

Marble rocks, so big that 20 pairs of oxen would not have been able to move them, lay thrown upon the shore at the root of Hoburgen, on the west side; they consisted of white and reddish grains, of the same kind as on the Karlsö islands and such as I had not seen on the island; the Karlsö islands can be seen in the distance from this side of the Hoburg, but how the water could roll these rocks such a long way I cannot understand.

* The top of the Hoburg is the type-locality for *Phalaena wilkella* L., *Syst. Nat.* 10th ed. 1: 541 no. 293 (1758); cf. *Fauna Suec.* 277 no. 899 (1746).—*W.T.S.*

Cancer macrourus, thorace articulato, caeruleus [*C. locusta* L.† = *Gammarus locusta* (L.)], found in the water on the shore, looked like reddish minute shrimps; it had four antennae; every antenna had 3 heavy joints at the base and then small ones, so very many that they could not be counted; the forehead was not pointed. At first were two pairs of feet with *chelae*, the uppermost parts of which had a flexible finger, but nothing to hold against; then were two pairs of legs with knees pointing backwards and feet forwards; then followed 3 more pairs of legs with knees pointing forwards, which seemed to correspond to those legs with which other crayfish fasten their eggs, and finally under the tail were two pairs of legs with two toes on each; the tail also consisted of two similar legs that were shorter, deeply cloven but flatter than the other ones; the dorsal claw was the shortest; on the tail there were two small scales; the eyes were white, punctate and shaped like a half-moon. The body consisted of 14 joints and the breast-shield was little bigger than the other parts.

Small *Neritae* [species of *Nerita, Neritina* and *Natica*] of different colours were in quantity on the western shore; they had a kind of lid on the mouth and two bristles, like two feet, on each side.

Cochlea testa pellucida, anfractibus quatuor, rictu ovato amplo, superficie rugis elevatis [*Helix balthica* L.] was also lying in the water by the shore; the animal was black and had two antennae looking like flat pointed ears.

There were many kinds of *Confervae* here in the sea. No. 1 [*Fl. Suec.* no. 1032] was much branched, thin and looking like wool. Every branch was distinct and water-coloured [transparent ?] with closely spaced red joints, which coloured the plant purple. No. 2 [*Fl. Suec.* no. 1031, *C. polymorpha* L.] ‡ was a green *conferva* like wool with many hardly visible joints; at the lowest branches there was a covering of small slippery grains. No. 3 [*Fl. Suec.* no. 1029] was very long and slender, with undivided branches which were pale and had no side branches; it looked like fine threads or the lichen (*usnea*) on trees. No. 4 was very branched and purple-coloured; the branches were arranged like bristles and looked as if they were cut off at the top; it was not as red as the first one; it was granular and eight times as thick; its joints were not as clearly discernible, since both the joints and the rest of the plant were brown, although not so deep brown as the first. No. 5 was much branched and very downy; the branches were yellowish and rather thick, but so densely covered with hair that one could not see them well.

Ulva tubulosa ramosa compressa [*Ulva compressa* L. = *Enteromorpha compressa* (L.) Grev.] lay abundantly on the shore where it had been thrown up; it was uneven, with many loops and cavities, and simple branches; it was also pale, not green, and often not thicker than a thread.

A self-eroded [i.e. weathered] stone lay a ¼ mile north of Hoburgen, so big that no human power could have brought it there; it was lying in the middle of the land, between the east and west shores. This kind of stone was altogether foreign to this country, hence the farmers showed it to us, since they supposed it would contain ore. The height of the stone was 3 ells [6 feet, 1.80 m], the breadth 7 ells [14 feet, 4.2 m] but hardly a fathom [6 feet, 1.8 m] thick between its west and east sides. It lay on the surface of the ground and

† The shore near the Hoburg should probably be designated as the restricted type-locality of *Cancer locusta* L., *Syst. Nat.* 10th ed. 1: 634 (1758); cf. *Fauna Suec.* 360 no. 1254 (1746).—*W.T.S.*

‡ Apparently *Pilayella littoralis* (L.) Kjellm.

consisted of a red spar like an Åland stone; between the cubes there was a black shimmer, here and there shining like gold in the sunshine. The sun, which constantly shone on the south side of the stone, made it all soft, so that it eroded itself and the gravel fallen from it lay around it as on a burial place, especially at the south side; this is consequently no ore, but a weathering stone. Since all *Saxa* or greystones are generated under the earth's surface and this stone was lying altogether on the surface, where there is no higher ground and all around is limestone, and since this kind of stone is foreign here, it is difficult to understand how it has arrived here, and if water had the power to roll it here from Sweden or Russia at the time this land lay under water.*

The burning midday heat was cooled by thin clouds, through openings in which the pale rays of the sun shone, and the people said that the sun was drinking water, and that it would rain tomorrow.

Sundre church, in the parish of Vamlingebo, was the southernmost church on Gotland, and lay a ½ mile north of Hoburgen; the meadows around this had some deciduous trees. We read four runic stones here, of which No. 1 was in the chancel and was nicely decorated with leaves skilfully engraved, No. 2 was outside the chancel, No. 3 at the south wall of the church and No. 4 north-east of the church.

"Castell" [citadel] was the name for a tall cylindrical wall near the churchyard, of which the thickness was 4 ells [8 feet, 2.4 m], the height 7 fathoms [42 feet, 11.6 m], inner diameter 4 fathoms [24 feet, 7.2 m]. The lowest part, which was roofed with a beautiful vault, was about 7 fathoms high; there was no door on this citadel, but on the north side was a big window high up in the wall. There were also small openings, unsuitable for guns. In the basement there was an opening in the middle of the wall, widening inwards, in which began a staircase, which between the *duplicature* of the wall [i.e. projecting pieces on each side for steps] led to the top of the citadel. This citadel was still rather well preserved, although it was used as a barn. We have seen similar citadels near the churches of Hambre, Öja and Fide, but they were more completely destroyed than this one. People here say that the citadels are older than the churches, but this seems strange to me, as they are always built quite close to them.

We saw stone houses with vaults and heavy walls on the farms; one may surmise that all these stone houses have once been dwelling-houses, or perhaps all farmhouses here were formerly built of stone; it would be desirable for the farmers of our times to do the same in a place so poor in wood but so rich in limestone.

In the afternoon we passed a little pine forest south of Vamlingebo, where we returned to spend the night.

There were hedgehogs [*Erinaceus europaeus* L.] here, and we caught a couple. The whole upper part of the body was thickly covered with spines; these spines were light grey or white encircled with distinct small dark rings, of which the outermost was the largest and darkest. The rest of the body and the head except for the crown was covered with light grey hair, as on a pig. The ears were rounded; the eyes small and black; the nose pointed; the nostrils had on the outside a little wavy lobe. The forepaws were like the paws of a bear,

* This stone was evidently a glacial relict.—*W.T.S.*

with five clubbed toes. The hind paws also had five toes, of which the second and the "thumb" [big toe] were the longest. The female had 8 teats, the male a *praeputium propendens*; both had a musk-like smell, and their excretions infected everything with the same smell. The male and the female hunted together at night, and they were most often seen in the evening. They build their nests in juniper bushes, ½ an ell [1 foot, 30 cm] above the ground, which nests are made of moss (*Hypnum*), rounded and hollow, like a squirrel's nest, and they are said to have usually four young. We had a remarkable proof of the fact that hedgehogs do eat meat; one of our hedgehogs devoured all the thrush described earlier, and left nothing but the thickest feathers; its excrements after this were like cardboard from the undigested feathers. When we let it loose on the shore, it proved that it was not afraid of water either; it went down to the sea at once and started to swim.

July 10 [p. 265]

Now that we had seen the southernmost cape of the land, we now returned northwards from Vamlingebo, but this time along the western shore; thus our road today was to the north-west along the seashore.

Bursvik, a big and long bay, cut into the land; from here is shipped all the sandstone of which the castle in Stockholm and other palaces are built.

The quarries or "The Old Dens", as well as other large quarries, were seen; they were not very deep; we counted the strata that were seen in most of the quarries in the following way:

1. "Aur" (*Gravel*) 8 quarter-ells [4 feet, 120 cm], which consisted of a little black mould but then some gravel with rubble and small stones of limestone.
2. "Kalkhäll (*Limestone*) 8 quarter-ells [4 feet, 120 cm], pale, rather slaty with undiscernible structure, from which formerly a good lime was burnt.
3. "Grushäll" (*Sandstone*) 4 quarter-ells [2 feet, 60 cm], which consisted of a sandstone distinguishable from the true sandstone in being drier and cracking more obliquely, hence it cannot be used.
4. "Ler" (*Clay*) 3 quarter-ells [18 inches, 45 cm], which is dry and disintegrates like marl between the teeth, is also somewhat slaty and it has a fine shimmer, like sandstone when it cracks; one might have believed that the sandstone was generated from this clay if there had not been a well-defined margin between the two layers. When this clay rubs against and lies on clothes which have grease spots, it removes these.
5. "Grushäll" (*Sandstone*) 1 quarter-ell [6 inches, 15 cm], like No. 3; on the surface of it there was sometimes a *crusta pyriticosa,* but not deeper down in the stone.
6. "Lera" (*Clay*) 1 quarter-ell [6 inches, 15 cm], like No. 4.
7. "Grushäll" (*Sandstone*) 4 quarter-ells [2 feet, 60 cm], like Nos 3, 5.
8. "Lera" (*Clay*) 2 quarter-ells [1 foot, 30 cm], like Nos 4, 6.
9. "God Sandsten" (*Good sandstone*) 4 quarter-ells [2 feet, 60 cm], which is quarried and hewn.
10. "Lera" (*Clay*) some inches, like Nos 4, 6, 8.
11. "God Sandsten" (*Good sandstone*) some ells, like No. 9.

The workmen do not go deeper down because of the water, although deeper down the good sandstone is both more abundant and better; if there were a hauling plant or a water pump here the water could easily be removed, since there is not much of it.

The gravel in the uppermost stratum (No. 1 above) is a kind of stone that lithologists call *Oolithum,* since it consists of hail-like grains, with one crust inside the other, like a crayfish stone or a sugar ball. Those who believe this *Oolithus* to be nothing else but petrified spawn could have an opportunity to see here more fossilized spawn than there has ever been real spawn in the whole world.

We found no filtering-stone, but I do not doubt that it could be found here; filtering-stone is nothing but a porous sandstone, which lets water through and clears filthy, rotten and salt water. All the good sandstone is wet all the way through as long as it is in the quarry; thus it absorbs water, which can pass through it, so that if it were only more porous it would be a filtering-stone. It is remarkable that the sandstone in the quarry is all wet, although it lies several ells above the water. This is the reason why dust from the stone does not injure the lungs of the workers here, as it does in other places, where it is hewn when dry.

The sandstone in the quarry disintegrates in perpendicular cracks that go *lineam rectam in infinitum*; the cracks parallel to the surface are called longitudinal, those that lead into the stone are called transverse, the latter ones do not open up, and there is no other matter in them. When quarrying is started, a pair of quarry-men go down together till they find the first longitudinal crack; then they part and follow the crack in different directions; the cracks go through all the upper layers, which must also be removed; if the quarry-men find large broad pieces of stone, they are very happy and *vice versa.* After these come others following other cracks. Since all the upper layers have been deposited inwards, towards the land, those who come last are rather busy removing all the waste left by their predecessors. The stone is quarried with hammers and wedges. Stones which are hewn to a definite size as ordered are called "Måttstenar" and cost more than others which are taken as they come and are called "Blåckstörer".

The price of the stone is calculated in square feet, e.g. 2 feet make one ell, every "quarter" consists of 5 inches, every foot is paid with 4 "styver", 3 21/25 "penningar", that is 13 "Daler Sm:t" for a hundred feet—and of course more than one stone makes up these hundred feet. If a stone is 30-50 feet it is paid with 5 "styver" a foot, from 50-60 6 "styver", up to 70 feet 8 "öre", up to 90 9 "öre", 90-100 feet 10 "öre" and so on.

There is an inspector from the government here every year from *medio Martii* to *medium Octobris* to watch over the quarrying for the royal castle.

Coffins of sandstone, as well as lids for them, hewn out of the stone, were seen here. Mostly ordered for nobles in Denmark, such coffins often cost 100 Daler Sm:t. This kind of work is easily ruined if rain falls on the sandstone and there is frost afterwards, since the stone will then crack; this happens quite often when the stones recently taken from the quarry and still humid encounter intense cold.

The Länsman [the chief of the local police forces], Jöns Winter, told me about an experiment concerning the water absorption of sandstone. A sandstone was hewn to one cubic foot, put into an oven and well dried. Its weight was then 9 lisp:d*. The same stone was then put into water, and when it was taken up it weighed 19 marks more and had thus absorbed 1/10 of its own weight. Thus it makes a big difference if a ship is loaded with wet or dry sandstone.

Sandstone is not used for millstones here, although it is here both abundant and hard enough, especially in the lowermost layers; the deeper down it lies, the harder and denser it is; instead, greystone with spar and quartz, which is common in Sweden, is used. Our Swedish millstones are often bought from remote places in order to get pure sandstone, which is quickly abraded and must be cleaned every other or third day, and the rubbed-off sand dust mixes with the flour, which has the same effect on our stomachs as if sand was mixed with the flour, i.e. it remains a long time in the stomach, produces obstructions in the mesenterium and has, in all, the same effect as a *poudre de succession* [a slow-killing poison], that is, it kills us silently. Millstones made of greystone do not need cleaning more than once a year, since it is so little worn.

We wished to see Näset, the cape which was on the north side of the inlet of Bursviken, and some foreign sailors took us there. Here grew *Arenaria maritima Rupp.* [probably *Spergularia marina* (L.) Griseb.], *Potamogeton foliis natantibus* [*P. natans* L.] and *Onopordon* [*Onopordum acanthium* L.]. On the beach lay all kinds of petrifactions, especially *Conchae striatae* and *Entrochi.* Here there were big fields, but the crops were thin, and the fields were neither manured nor drained.

"Oxel" [*Sorbus intermedia* (Ehrh.) Pers., whitebeam] was the only tree here.

Salicornia herbacea grew over the whole shore where water had stood during the winter, but the sheep had eaten so much of this salty herb that it was difficult to find it without looking carefully; we were delighted to find this plant in Sweden, as the true Spanish soda, so indispensable for glassworks, is made from it by burning. If anyone should want a large quantity of it, the only thing to do would be to keep the cattle away from the shore.

I know well that burning soda out of seaweed or *Fucus* has been tried, but the soda is not so good as that from *Salicornia* or *Kali.*

More than 20 seals looked up from the water one after the other as we crossed. This headland is the best hunting ground for seals on Gotland, especially on a sandbank shaped like a half-moon with an inlet of shallow water, where the seals go to rest. The inhabitants place their nets in parallel rows outside the opening one behind the other, then they drive the seals from the curved sandbank, so that they are caught in the nets. The people always call the seal's fore feet its hands. The seal has a very sensitive nose, and a very light blow there kills it.

We saw a neat way of killing hares; the hare is lifted by its back feet and is struck with the hand over its neck, which kills it immediately.

---

* The Linnaean abbreviation "lisp:d" refers to the lispund, Liespfund, lispound or Livonian pound long used as a measure of weight in Baltic countries, as also in Orkney and Shetland, but varying from place to place. Here it is probably equivalent to about 19 lb avoirdupois (8.5 kg) although it ranged up to 30 lb. The mark was a lesser measure of weight, equivalent here to about 1 lb avoirdupois, being in Sweden 1/20th of a lispund, although in some places 1/24th. Schreber in 1764 translated the Swedish "Mark" into German as "Pfund".—*W. T.S.*

Gryllus capite thorace elytrisque superne viridibus [*G. viridulus* L.* = *Stenobothrus viridulus* (L.)], or small grasshoppers, were here on the headland in such abundance as to be like newly sown seed under our feet; head and breast were dark green above, wings light green but the outer had a white margin on the outer side; abdomen, thighs, breast and face were pale; on some of them there was a black stripe on the breast on each side from the eyes to the wings; on some there was a dark line instead of the white one on the wings, but they were all of the same species.

Cimex ovatus exalbidus capite acuminato [identity uncertain] was found in the grass. It was completely pale, with a few longitudinal lines over head and breast that were still paler than the rest of the body.

Chrysomela atra, punctis excavatis contiguis [*C. tanaceti* L. = *Galeruca tanaceti* (L.)], an insect that we had only seen on Heligholmen before, was here in great abundance. It crept over the ground very slowly, and the elytra had a protruding margin on the sides [see p. 13].

"Swärtor" [*Anas fusca* L. = *Melanitta fusca* (L.), velvet scoter] of both sexes were seen, and we learned how to distinguish them. The male was all black, and the wing-coverts had a white spot; the female had almost the same colour, but less dark; the male had a white spot at the ears and the female a light grey spot. The male had a yellow bill with a black knob at the base, but the female's bill was black without any knob.

"Pracka" was the local name for the *Mergus* [*M. merganser* L., goosander], which we described earlier (*It. Oel. 49*, [entry of 2 June]). Here we had the opportunity to see both sexes. The male had a black head sprinkled with rusty spots. The sides of the abdomen were white with black wavy markings; the breast greyish with black spots. The back black, the pinions white, wing-coverts white, but the lowest were black at the base, which made a white spot with a transverse black line appear on the wing. The abdomen white; the tail dark; the crest at the neck hanging down. The bill black with sharp teeth pointed backwards. The female was somewhat dark, with rust-coloured neck, white abdomen, and also had a crest hanging down.

The evening came, a pelting storm blew up from the north-east; the voyage around the bay would have been 2 miles; thus we tacked back. The waves soaked our clothes, but we managed to cross and found nightquarters at the home of the länsman, Jöns Winter, in Botveda.

July 11 [p. 271]

This day we used mainly for our own mending, since our clothes were torn by bushes, rocks and much riding.

The sheep in this district do not graze in herds as in other places, but never go more than 4 or 5 together; if they were to gather in bigger flocks, they would never have enough to eat on this poor soil; these sheep, which are left in the open all year around without shelter at night, are herded together once or twice a year, when all the inhabitants take part, at Bartholomew time [24 August], when the young sheep are sheared, and at Matthew time [21

* Bursvik, Gotland is the restricted type-locality for this species; cf. L., *Fauna Suec.* 198 no. 626 (1746).–*W.T.S.*

September], when the old ones are sheared and picked out for slaughter or sale; the sheep are recognized from the ear markings.

One may wonder why hedges are not grown here, where the hazel grows all over the meadows, as well as hawthorn, and fences cost so much.

Fences are brought from places three miles away. 12 wattles make a "Värp" [pile] that costs 3 styver. Twelve posts make a "börd" [burden] that costs 1 styver. There are 24 "bördor" on a "lass" [load], thus one "lass" costs 24 styver. [*According to*] *J. Winter.*

The meadows were full of *Absinthium maritimum* [*Artemisia maritima* L.], *Scutellaria foliis integerrimis* [*S. hastifolia* L.], *Anagallis flore rubro* [*A. arvensis* L.]. In the fields grew *Cotula foetida* [*Anthemis cotula* L.], *Delphinium* [*D. consolida* L.] and *Ranunculus echinatus* [*R. arvensis* L.].

Anthyllis [*A. vulneraria* L.] varied delightfully here in several colours: *flore luteo, flore albo, flore ferrugineo, flore coccineo.* Together with this grew *Ononis* [*O. repens* L.], here called "puktorne".

A plant growing here seemed to be *Jacea nigra pratensis latifolia C.B.* [*Centaurea jacea* L.], but it was impossible to know for sure, since the flowers had not yet opened, and the stalks were softer than usual. The involucre, which was evident, made us doubtful as to the species, because the scales were oblong and torn, and the leaves were downy and silver-coloured.

Wild carrots [*Daucus carota* L.] grew here as big and tall as dill. The root was white and smelled like ordinary carrot. The stalk was thinner and grooved. *Umbella alba; Involucrum universale pennatifidum, partiale integerrimum: Semina hispida.* This plant should be examined more closely when the fruits are ripe.

Aster pratensis autumnalis conyzae folio, Flor. Svec. 694 [*A. 694 dysentericus* of Index = *Pulicaria dysenterica* (L.) Bernh., but Linnaeus's plant was probably *Inula britannica* L.], never seen growing wild in Sweden before, was found in the ditches between the fields. The root was creeping. The stalk a quarter-ell [6 inches, 15 cm] tall, straight, reddish, covered with white hairs. The leaves lanceolate, amplexicaul, downy, soft, serrated with reddish teeth. Involucre many-leaved, downy, with smaller leaves inside. Florets yellow, with narrow rays in the circumference. Receptacle naked. Seeds flying with down.

The blackthorn bushes [*Prunus spinosa* L.] were all bare, covered with silk-like cobwebs, in which hung thousands of pale pupae with black breasts and black-tipped abdomens.

Cerambyx cinereus, elytris punctis nigris fasciaq. nigra, antennis corpore sesquilongioribus [*C. nebulosus* L.* = *Leiopus nebulosus* (L.)]. This was a little iron-grey insect with black spots, denser at the base of the wings; it had a black band across the middle of the elytra.

Phalaena pectinicornis elinguis, alis deflexis atris: superioribus rivulis flavis, inferioribus rubris: maculis nigris [*P. plantaginis* L.† = *Parasemia plantaginis* (L.)], a beautiful but not common moth, with thin antennae, black wings with pale yellow veins, but the under wings red with black spots. The body was

---

* "Gotlandia ad fodinas Burvicenses" is the type-locality of *Cerambyx nebulosus* L., *Syst. Nat.* 10th ed. 1: 391 (1768); cf. *Fauna Suec.* 163 no. 483 (1746), 2nd ed. 187 no. 650 (1761).—*W.T.S.*

† Bursvik can be accepted as the restricted type-locality of *Phalaena plantaginis* L., *Syst. Nat.* 10th ed. 1: 501 (1758).—*W.T.S.*

black on the back, and purple on the sides with black spots; a purple ring encircled the neck.

We left Bursvik at 7 p.m. The road was very straight. The land began to be covered with forests of birch, hazel and oak. On our left we had the vast Bursvik bay quite close. On our right we saw the churches of Öja, then Fide and lastly Grottlingebo.

We passed Näs church, in the parish of Habdum, on our left, after ¾ of a mile's journey. Opposite the church was a vast field almost covered with limy clay. The soil was all bare, cracked and rough, like dough that has been left to rise too long. On this grew little more than sheep's fescue [*Festuca ovina* L.] which was so sparse that there was more than a quarter-ell [6 inches, 15 cm] between the plants, and *Lappula* [*Myosotis lappula* L. = *Lappula myosotis* Moench], more than in any other place, but never more than a finger tall. There were also a few junipers and a little *Serpyllum* [*Thymus serpyllum* L.], *Sedum flore albo* [*S. album* L.] and *Carlina sylvestris vulgaris. Clus.* [*C. vulgaris* L.]. Around here pine and juniper forest covered the ground, and we lost sight of Bursvik and the southern plain.

Habdum church is situated 1½ miles from Bursvik. We arrived there at 9 p.m. Although it was getting dark, we could read some runic stones in the churchyard, one of them with deeply cut letters. We had planned to stay here during the night, but a malignant burning fever had afflicted some people at the farm; thus we went on.

The rakes used for hay were made in a strange way here, the teeth of the rake being attached to it *ad angulum acutum.*

Sempervivum [*S. tectorum* L.] grew here on the roofs, preventing the turf from falling down.

We left Alva church on our right.

Darkness came upon us, so that we saw nothing except that the cats caught cockchafers and ate them. At midnight we arrived at Fardum.

July 12 [p. 275]

Fardum church kept us here till the afternoon: we attended the service and heard the Royal Edict against the harmful credit trade [see entry of 2 July] being read; it will prove difficult to stop this, if another little town is not founded, for instance at Slite or thereabouts.

Hops are of two kinds, male and female. The female is the real hop, the buds of which are used for brewing. The male is called here "Fuk-humble" [sterile hops], in Sweden "Gall-humble" or "Frö-humble", for it has no buds, only many small flowers. There was a hop-yard of a considerable size here, but there were only sterile hops in it, and thus it did not yield any fruit. The owner had asked advice from old and wise farmers, who had told him that when the roots of the hops grow old and get a thick bark, they only yield sterile hops, and that there was nothing to be done but to dig up the yard and replant it with the same roots, which would then give new fresh roots and good fruit. This was done, but the plants that grew up from the replanted roots still gave sterile hops, and the owner waited in vain for the real hops. The real hops will never change their roots to give sterile hops, nor will the sterile hops give any fruit, however it is replanted. The sterile hops come into the hop-yard when the hops drop their seeds, which give rise to male and female plants, in the same way as eggs from a hen give rise to cocks and hens; the sterile hops grow better and the female hops are easily suffocated, which can spoil the best hop-yard. The females bear fruit without the males, since the hop cones are not real fruits but the calyces or sepals; there will be no seeds, unless the pollen of the sterile hops is blown on to the hop cones. It is true that the hops give fruit without the males, but it is also true that if they are fertilized by the males, the hop cones get better, but then the owner can never be sure that he will not get sterile hops in his yard.

Beehives were not seen here, although there were so many on Öland. There was no heather here; the honey would be much whiter if bees were brought here. We were told that the late bishop Esberg had brought bees here formerly, but they had not survived, which should not be regarded as a precedent, since he lived in Visby, where the bees are exposed to a strong wind each day and the swarms have the opportunity to escape to the many old churches and towers.

We saw no flax on Gotland, and when we asked the people why they did not cultivate such an extremely useful plant, they answered that they had often sown a whole pailful (2 bushels) of flax seed in excellently prepared and manured soil, but got no more than a pound of flax, although it grew to more than one ell [2 feet, 60 cm] high; therefore they preferred to sow grain and sell it, and buy their flax for the money thus received, which will give them 3 or 4 times as much of it. One could see from this that the farmers were very unaccustomed to this kind of cultivation.

The houses which the farmers built here were generally of boards, trimmed into the beams; 3-4 boards are hewn from each log, which gives a board that is 3-4 inches broad and makes the walls look very even, and they are often smoothed. The upper edge of the boards is tarred before the moss is added, which makes the joints very tight. The farmers use neither double floors nor fillings in their houses: those who do not need dampers for the fire do not care much about warmth.

The Gotland cellars are mostly unusable in summer, except in Visby; the beer gets sour in them, which is supposed to depend on the limestone, which absorbs water especially in humid weather.

In old corves* the fish do well, but in new ones they die; it was said that if the new ones are smoked with straw the fish will do well in them too, which could be easily tested.

Many worms, looking like piles [*Oestrus haemorrhoidalis* L. = *Gasterophilus haemorrhoidalis* (L.)], were fastened in the anus of the horses, a condition which *Réaumur*, [*Mém. Hist. Insectes*] *Tom. 5. Tab. 35. Fol. 1, 2, 3, 4, 5* has illustrated. The farmers told us that these worms are deposited in the noses of the horses while they are standing in the sun. This makes them sneeze and snort, until the muzzle almost bursts, and the worms advance through the gut; and after one year they reach *intestinum rectum vel anum.* I leave others to judge how certain this is. I know well that the muzzle worm of reindeer and sheep is another species, which has also been depicted by *Réaumur. Tom. V. tab. 35. Fig. 21, 22, 23, 24,* but they are closely related and of the same genus

---

* A corf is a large floating basket or cage in which caught fish or lobsters are kept alive in water.—*W.T.S.*

as the gadfly or *e genere Oestri.* Perhaps they could be used in the treatment of piles as well as leeches?

We spent the night at Livoistad [Loijsta].

July 13 [p. 278]

The silentness [i.e. lack of news] here in the countryside is very great; newspapers do not come often, and except for two judges and one captain, there were no persons of standing on all Gotland, if those living in Visby are excluded: the nobility have no estates here, and the farmers gratefully serve "the shepherds of their souls" [i.e. the clergy], helping them with the farming, with ploughing, manuring, mowing, harvesting, cleaning, driving, building, etc.

At Eksta church there were two runic stones in the churchyard, one outside the church door, 2 fathoms [12 feet] broad and 1 fathom [6 feet] long, the other to the east of the church.

After ½ mile's journey we arrived at the shore and went on board in the company of Judge Laurin, who was the owner of the Karlsöen; after ½ mile's sailing, against the wind, we arrived at Stora Karlsö.

The two Karlsö islands, the big one and the small one, are ¼ of a mile apart and less than ½ a mile from the mainland; both of them are very high and very steep on all sides; resting on solid rock and surrounded by the sea, like Torsburg and Hoburg on Gotland, which mountains must once have been similar to the two Karlsö islands at a time when Gotland was under water.

Stora Karlsö had on the west side some huts for fishermen and also six cylindrical stone houses, that had formerly been *Coupées* or small fortifications. Close to the fishermen's huts was a big field with an upright stone, like a runic stone, with an inscription* on the east side:

JOHANNES DIETERICUS GRONHAGEN
L.B. Gotlandiae Gubernator
Concomitantibus
Conjuge & quatuor Liberis

PATRICKIO KRABBE
Provinciae Camerario
&
ERICO FIELSTRÖM
Districtus meridionalis Praetori
Ob felix susceftum iter ad utrasq;
Caroli insulas
Diebus XII & XII Junii A.C.
M.DCC.XXXII.
In declarandam gratitudinem suam erga
Deus
Et in memoriam relinquendam
Posteris
Hoc monumentum
Propriis impensis poni curavit.

* This commemorates the visit in June 1732 of Johan Didrik Grönhagen governor of Gotland, accompanied by his wife, his four children, Patrik Krabbe and Eric Fiellström, to the two Karlsöarna.—*W. T. S.*

On the west side of the stone the following words were written [in Swedish]:

As a token that he has
Been on Stora Carlsö
The governor and father of the land
After visiting Lilla Carlsö
Has with his own hand
Had this stone erected here
It is taken from his land
And can tell others about it.

Carlsö, d. 13 Junii, 1732, J.G.N.

A rather big heap of stones was seen at the south cape of the island; we searched very closely but in vain for runic stones, and the inhabitants could not show us any. The stone heap had a hole at the centre; when we turned over some stones in it, we saw a great many very rare worms or *juli.*

Scolopendra teres pedibus utrinque CXX [*Julus sabulosus* L.*] was 2 inches long, thick as the quill of a dove, smooth, grey with two pale yellow and even longitudinal stripes on the back; the body consisted of about 60 joints and every joint was grooved and paler at the margin. The feet were many, thin as hairs, light grey; the antennae consisted of five joints; as soon as the worm was touched, it coiled like a spring, lying on its side with its head in the centre.

The hedgehog, which we had brought with us, ate grasshoppers hopping in the field with an excellent appetite, as well as the cockchafers that appeared in the evening and some snails in the grass.

We found springs here, not far from the huts, with water which was much praised, and tasted clean, but I did not dare drink it because of the worms on the bottom.

These worms, *Gordius* or *Seta aquatica* [*Gordius aquaticus* L.], look like a horse-hair in shape and size and are pale with black head and tail. When these worms are cut in many pieces, every piece moves and the farmers in Småland told us that if these pieces are thrown into the water, every one regenerates into a whole animal, which would seem incredible if it were not known that pieces of tapeworm can grow to full-sized animals after many fathoms of the worm have been torn off. Moreover, this worm must, like all other animals, have eggs and seeds that develop into full-grown animals when they enter man's body, in the same manner as tapeworms, earthworms and other small worms which grow in the body of man, when the eggs have entered it from water. A fly that lays its eggs in milk, or in any food, never gives rise to intestinal worms, least of all the housefly, which is viviparous. If the eggs of the *Gordius* enter the stomach with the water and are hatched there, the worms could easily bore through gut, liver, lungs and heart and cause strange and unknown diseases, as well as *Vena medinense* or *Dracunculus persarum.* Perhaps water drinkers should be more careful, especially here where the water contains lime and this creature thrives so well. Perhaps some of the many intestinal troubles here are

† Stora Karlsö is the restricted type-locality for *Julus sabulosus*: L., *Fauna Suec.* 362 no. 1262 (1746).—*W.T.S.*

due to *Gordius.* And perhaps the interminable colicky pains of the Lapps, who drink their water from cold springs over long periods of time, are also caused by this animal.

Madreporae, or star corals, covered all the shores, and sometimes pyrite particles were found in them; they were also seen in the strata of the rock on the island, where it was perpendicular. I will not enumerate the corals since all the Gotland corals that have been found during the journey are recently published at Uppsala in a dissertation under my *praesidium* by auscultant *Henr. Fougt* [*Dissertatio Corallia Baltica adumbrans . . . subm. H. Fougt,* 1745; cf. A. G. Nathorst, *Carl von Linné als Geolog,* 48, t. 1: 1909].

The stone used for the churches and their marble pillars, which we have seen in Visby as well as in the countryside, is said to be taken from Karlsö; the bedrock here consists of grey and reddish marble.

During the night we rested in a fisherman's hut on Stora Karlsö; no people live on either the bigger or the smaller island.

July 14 [p. 282]

Stora Karlsö was very high, except for the south cape, with a horizontal infertile level on top, similar to the *alvar*-land on Öland. There was a heap of stones on top of it, 1 fathom [6 feet] high, upon which grew an ash tree, 2 fathoms [12 feet] tall and as broad in the crown; this was the only tree on Stora Karlsö, and it was used as a landmark by the seafarers.

Saponaria calycibus pentaphyllis, corymbis fastigiatis, foliis linearibus, caule adscendente. Flor. Svecic. 346 [*S.* 346 *carolina* of Index = *Gypsophila fastigiata* L.], a plant that I have not hitherto seen either in Sweden or in any botanical garden, grew everywhere in the infertile field. This herb has been observed by botanists, but it has not been well described and hence has been confused with other plants. The first to see and describe it was Thalius in *Hercynia p. 115* [J. Thal, *Sylva Hercynia*; 1588], where he calls it *Symphytum petraeum s. Gypsophyton minus.* The second was Mentzelius in *Pugillo* [C. Mentzel, *Pugillus Plantarum rariorum*; 1682] who has given a short description and a rather good drawing, *Tab. 2,* under the name of *Polygonum majus erectum angustifolium, flore candido,* but he has made the leaves too big. Finally it has been found by Scheuchzer in the Swiss Alps, but he has confused it with another rather similar plant. I will give its description here in order to make this clear to botanists. The root is perennial, as thick as a goose quill, and goes rather deep. The radical leaves form a tuft, as usual in pinks; they are linear, pressed down and somewhat thick. The stalk, 2 quarter-ells [1 foot, 30 cm] high, at first lies on the ground but rises up when the flowers appear; it is divided into 6-8 reddish joints, round and smooth, and ultimately becomes wholly red. The stem leaves are similar to the radical leaves but rather smaller, smooth, flat above, keeled below. The inflorescence terminates the stem and looks like a brush (*Corymbus fasciculatus, fastigiatus, dichotomus*) composed of flowers, of which the calyx is bell-shaped and divided to half-way down into five red lobes. The petals are five, broad, oblong, blunt, outspread. The filaments are 10, white, as long as the petals; after one day five bend back and next day the other five also bend back. The anthers are also white. The ovary is globular. Styles two, thread-like, as long as the stamens. The figure that

Plukenet gives in his *Phytographia Tab. 37. Fig. 1* [1691] looks rather similar to this plant.

The other plants common in this high field were *Anthyllis* [*A. vulneraria* L.], *Asperula rubia cynanchica dicta* [*A. tinctoria* L. = *Galium triandrum* Hylander], *Galium luteum C.B.* [*G. verum* L.], *Galium album quadrifolium erectum Celsi* [*G. boreale* L.], *Linaria* [*L. vulgaris* Mill.], *Asclepias* [*A. vincetoxicum* L. = *Cynanchum vincetoxicum* (L.) Pers. = *Vincetoxicum hirundinaria* Medicus], *Scabiosa capitulo globoso minor C.B.* [*S. columbaria* L.], *Serpyllum* [*Thymus serpyllum* L.], *Helianthemum* [*H. chamaecistus* Miller], *Geranium pedunculis simplicibus unifloris* [*G. sanguineum* L.], *Erigeron vulgare Fl. Lapp.* [*E. acer* L.], *Solidago, quae virga aurea* [*S. virgaurea* L.] and *Origanum* [*O. vulgare* L.], especially where the rocks were steep, in such numbers that the dyeworks could hardly have a better supply of this anywhere else. *Adonis radice perennis* [*A. vernalis* L.] was also common here.

The sheep grazed all day long on this dry field, on the barren rock, where the sheep's fescue grew sparsely and everything was so meagre that one could believe that even arctic lemmings would have died of starvation. It is amazing to see how Nature deals with the sheep; they go all day on the dry rock without once touching the tall grass that grows between the mountainside and the sea, and yet this grass was so tall that it reached our knees when we walked in it; in spite of this, the sheep are very fat. Our greatest art and our true science is to observe and investigate the laws of nature. The farmers shear these sheep only in autumn, at the feast of St. Bartholomew and St. Matthew [24 August and 21 September]; but in spring, when the sheep cast their wool, it is combed out with the hands.

Tjuvhålet [The Thief's Den] was the name of a cave in the west side of the mountain slope. It is reached from the edge of the plateau by a ladder that leads to a path on a little projecting rock, which finally becomes so narrow that only one man at a time can pass; on one side there is the rock wall, on the other the deep precipice; after a short passage a big cave is reached, which can accommodate some hundred men; it cannot be attacked either from above or from below, and if there were only one sentinel posted at the entrance, the cave could be well defended against a whole army, and that with only a stick to throw down anyone who tried to reach the cave, since there is nothing to hold on to.

Artemisia foliis pinnatis, ramis adscendentibus hirsutis, floribus globosis pendulis, receptaculo papposo, Fl. Svecic. 669 [*A. 669 rarissima* of Index = *A. rupestris* L.] was found on the northern part of Stora Karlsö, directly below the place where the ash tree stood. The plant was withered and similar to *Abrotanum campestre, Fl. Svec. 668* [*Artemisia campestris* L.], which grew here too; but when one examined it more closely, this was found to be the rarest of all the plants we saw during the journey, or maybe a completely new plant. The root was loosely attached to the soil and rather short, almost like southernwood. The stems were short; they lie on the ground until the flowers appear, when they straighten up; the whole plant was rather hairy. The leaves were short and feather-shaped (*pinnata*) with shorter veins at the base, while the other veins were linear and parallel. The flowers which arise from the base of the leaves [i.e. from the leaf-axils], especially on the upper part of the stem, are pendulous on long flower-stalks and almost always solitary; they are bigger than on any other *Artemisia.* A special distinguishing feature of this plant is that the flowers are big and globose, looking like round pills, which makes it easy to differentiate them from others. I brought it with me to the University Garden in Uppsala, where it has survived the winters and spreads by its roots [i.e. rhizomes] but, because it flowers late, it has ripened no seeds for several years [see p. 9].

From Stora Karlsö we went to Lilla Karlsö by boat. As soon as we came out on the water, gulls and other sea-birds started to fly about us with much screeching and screaming, as they did when we arrived yesterday. We saw gulls, scoters, razorbills and black guillemots in the air, and we shot some of them to investigate them more closely.

The razorbills (Torderna) flew rapidly around the boat several times without fear, and, although we shot at the birds, they did not go away but became ever more eager to fly above our heads. This bird is very rare in Sweden and is called *Alca, rostri sulcis quatuor, linea utrinque alba a rostro ad oculos, Anim. Svec. 120* [*Alca torda* L.]; it is big as a raven, heavy and fleshy, all black except for breast and abdomen as well as the lower wing coverts and the tail coverts, which are all white. The outer pinions of the wing are 11 and black, the nearer the wing tip the longer; the inner pinions are 15, of equal length, black with white tips. The tail feathers are 12, black and short, the central ones being slightly longer. The feet are black and webbed with three toes, of which the central one is the longest; there is no thumb. The bill is very strangely shaped; it is black, compressed, obtuse and raised at the middle, the upper jaw is bent like on a hawk, and has four broad transverse furrows, but the lower jaw has only two furrows; the second furrow on both jaws is white, which gives it the strange appearance. The mouth is very big and three times longer than the bill. The edges of the bill are very sharp, and in the palate there are a few small 'teeth'; the tongue is fleshy. From the root of the bill there is a white line over the forehead to the eye. The thighs are placed behind the *balance* of the body, as in a diver, because of which it swims and catches fish well, but it cannot walk like other birds in a horizontal position, it has to walk upright like a man; this makes it look very funny with its black coat and white vest, its wrinkled forehead and hooked nose, as if it was wearing spectacles. It builds its nest in the highest and steepest rock crevices, which are almost impossible for men to reach; it lays only one egg and, if this egg is removed, it immediately lays another one. If a person walks beneath the cliff and shouts loudly "Out there" or something like it, the birds come flying out in great numbers. If one of them is shot and injured, but not so seriously that it cannot reach its nest, it is thrown out by the other birds. If one shoots at a bird and misses, it does not escape but flies above the hunter, so that there is time to reload the gun and shoot once more; and one can shoot 3-4 times at the same bird. They fly very rapidly like a dove. They are away in winter, but return on the first day of May, and leave their summer quarters at the time of St. Laurence's festival [10 August]. The farmers told us that all the birds are present in the morning of St. Laurence's day but not in the afternoon, for then all the birds are gone and none returns until next year.

The black guillemot (Grautle) [*Uria grylle* (L.)] was a bird that we only saw flying; and it was rather similar to the razorbill but smaller. The farmers told us that it also walks upright, has red feet and is red under the tail. The breast is black, not white as on the razorbill. It was said to have two eggs and build its nest in the crevices, but never more than 2 fathoms [12 feet] above the water level; is it *Columbus pedibus tridactylis palmatis, Anim. Svec. 124?*

We also saw another bird flying that was rather similar to the razorbill, but the back was more greyish blue and the bill smaller, straighter and longer. According to the farmers, this bird also walked upright and laid one egg; is it *Alca, Anim. Svec. 188*? [*Alca arctica* L. ?].

A sparrow, that was almost all black, was seen on the north side of the island where this was steepest, but it was impossible to shoot it; perhaps it was a shearwater, *Procellaria, Anim. Svec. 249, act. Stockh. 1745, p. 93* [*Procellaria pelagica* L.].

Several gulls or *Lari majores* were shot; they were similar to the ones described earlier [see 30 June], but back and wings were dark grey. The inner pinions from 10 to 30 were equally long and black with white tips, but the outer ones increased in length towards the tip of the wing and the tip was less white, except the outermost or first, which was not white at the tip but had a white spot below the tip. The lower jaw had a saffron-yellow knob at the tip. This bird is *Larus albus, dorso cinero fusco, Anim. Svec. 126* [*L. fuscus* L., Lesser black-backed gull].

Grey gulls flew together with the above, but they were not as numerous; they were similar in shape and size, but the colour was quite different, like a snipe, a mixture of light grey, rusty brown and black; neck and breast were light grey, as were the undersides of the wings. The pinions were grey with white tips, but the outer ones had no white spot below the tip. The tail feathers were white and sprinkled with sooty black flecks, especially the middle one; towards the tip they were blacker but white at the tip itself. The legs were light grey and not yellow; the corners of the mouth were not red and there was no red ring around the eyelids as on the other gull, and the bill was quite black. When they were shot, they spewed up handfuls of small fish. Later on I fed such a bird with meat, from the autumn until next spring, and found out that it was nothing but the young of the other gull; in spring it became white with a grey back and quite similar to its parents.

The Lilla Karlsö island is situated ¼ of a mile from the big island, closer to Gotland; it resembles the other island with its steep sides of naked stone and its barren field on top, upon which were some trees, such as hawthorn, service and ash. On the southern side it was very green. *Cochlearia* [*C. danica* L.], *Melampyrum arvense Riv.* [*M. arvense* L.], *Astragalus leguminibus lunulatis biventricosis, caulibus procumbentibus* [*A.* 591 *procumbens* of Index = *A. glycyphyllos* L.], *Thalictrum minus C.B.* [=*Th. minus.* L. var. *kochii* (Fries) Neum.], *Caucalis Fl. Svec. 225* [*Torilis japonica* (Houtt.) DC.], *Athamanta Fl. Svec. 229* [*A. libanotis* L. = *Seseli libanotis* (L.) Koch], *Lithospermum officinarium Fl. Svec. 151* [*L. officinale* L.], *Isatis* [*I. tinctoria* L.], service tree and hawthorn grew here.

Lactuca foliis pinnato-sinuatis denticulatis acutis subtus laevibus, caule glabro. Fl. Svec. 646 [*L.* 646 *quercifolia* of Index = *L. quercina* L.] was found on the south-east side of Lilla Karlsö; I had never before had the opportunity to see this plant, and Ray is the only one who has described it so clearly, that one can be sure that he has such a plant; he calls it *Lactuca foliis quercis* in

Hist. Plant. 221. The root is fleshy and blunt. The stem is 1 ell [2 feet, 60 cm] long, straight, rounded and glabrous, without branches; it ends in an erect, narrow and long inflorescence. The leaves resemble those of a milk thistle [*Sonchus oleraceus* L.] with more pointed teeth and look as if gnawed. The calyx [involucre] is almost cylindrical, scaly, glabrous and sprinkled with rust-coloured spots; its scales are straight and provided on the back with an outward spreading scale [see p. 11].

From here we sailed to Klintehamn.

Klintehamn had more ships than we had seen in any other Gotland harbour; they were loading lime, beams and boards.

The beams are hewn by the farmers, who bring them home and trim them next day and take them to the harbour the day after; usually they get 10-14 styver a beam for work, horses and wood, but they neglect their farms, which suffer from it: the meadows are overgrown with bushes and are not mowed in time; yet the farmers complain of too little hay.

On our way we passed Sanda church; it was built of marble, with splendid pillars inside. In the graveyard lay a runic stone with the following inscription: Bolaider i Boligaby lit dinna Stain gera iftir Olof Fadur sin &c. Bidin firi Olaafs Sail i Boligaby, Gud gifi saili din &c. [Bolaider in Boligaby had this stone made for Olaf, his father. Pray for Olaf's soul in Boligaby, God give your soul &c.]

Ruins of an old four-cornered house or tower were seen not far from the church; in the basement there were two vaults but in the upper storey there was one vault with a beautiful pillar in the middle; all the openings in the basement were very narrow. On top of the vault grew *Epilobium floris difformibus, pistillo declinato* [*E.* 304 *irregulare* of Index = *E. angustifolium* L.]. The inhabitants believed that these stone houses had been built during the times when there were pirates in the Baltic, so that inhabitants could hide quickly with all their belongings.

This night we rested in Söderting.

July 15 [p. 291]

Mästerby church, in the parish of Sanda, was passed; there was a runic stone in the graveyard that we could read and another one that was illegible.

Dogwood (Benwed) or *Cornus foemina Fl. Svec. 131* [*C.* 131 *ossea* of Index = *C. sanguinea* L. = *Swida sanguinea* (L.) Opiz] grew everywhere in the scrub more commonly than in any other place and was often 3 fathoms [18 feet, 5.5 m] high; it had recently put forth its white flowers.

The fallows were yellow with *Buphthalmum* [*Anthemis tinctoria* L.], the balks of the fields were adorned by blue *Cichorium* [*C. intybus* L.] and the meadows shone with *Aster montanus luteus salicis glabro folio Fl. Svec. 695* [properly 696; *Inula salicina* L.].

Just before we arrived at the estates of Roma, we passed a fir tree 4 fathoms [24 feet, 7.3 m] high; it was naturally grown in the shape of a cone, as if it had been trimmed by a gardener, and the sky could not be seen anywhere through its branches.

In the afternoon, we arrived at Roma monastery, just before there began a thunderstorm with lightning and rain to soak the dry earth that had received no rain since we came to Gotland. We stayed here till the following afternoon.

July 16 [p. 291]

Roma monastery, which is now the residence of the District Governor [Jakob von] Hökerstedt, is in an incomparably pleasant place, and furthermore well built with beautiful stone houses and fine gardens. In the south there are vast meadows and in the north large fields, both surrounded by evergreen forests, which border on all this loveliness. In the north one can see the church of Roma from the residence on the north side of the fields; a straight main road leads there.

The barn is one of the grandest and one can see how time can make its Metamorphoses here greater than ever Ovid did: one sees a marvellous monastery, built of hewn and polished marble, transformed into a barn, which is thus the finest barn in Sweden.

The herbs here were *Lycopus* [*L. europaeus* L.], *Lythrum* [*L. salicaria* L.], *Utricularia* [*U. vulgaris* L.] and the rare *Marchantia calyce communi quinquefido hemisphaerico* [*M.* 932 *quinquefida* of Index = *M. hemisphaerica* L. = *Reboulia hemisphaerica* (L.) Raddi], which all grew near the marshes. *Sparganium* [*S. ramosum* L.] stood in the water and *Odontites* [*O. verna* (Bellardi) Dumort. ssp. *serotina* (Wettst.) E. F. Warb.] in the fields.

Ononis [*O. spinosa* L.] is called iron-root (Järnrot) here, since it is very troublesome to the farmers when ploughing, being so tough.

"Kaipe" or *Porrum capitulo bulboso erecto, foliis planis subcrenatis vaginis ancipitibus It. Oel. 60* [entry of 4 June], *Fl. Svec. 266* [*Allium scorodoprasum* L.] grew so abundantly in the meadows that it is no wonder that the milk as well as the butter tastes of garlic in spring.

Scordium [*Teucrium scordium* L.], a plant that is often prescribed in the pharmacies and imported annually, was found growing wild in the marshy meadows among grasses, on the north side, at the place where sedges begin to flourish.

Carduus nutans, Joh. Bauh. [& J. H. Cherler, *Historia Plantarum universalis*] *3. p. 56* [*C. nutans* L], a herb rarer in Sweden than in France, grew along the main road close to the monastery in abundance; it was like a common thistle [*Cirsium vulgare* (Savi) Ten.] but the stem is more wavy, with strong spines; the heads are not downy, but the scales are broader and more spreading.

We were asked to test the water in *Walliers* well at the monastery; it did not change with *Mercurio sublimato* [mercuric chloride], *Spiritu salis armoniaci* [spirit of sal-ammoniac, chloride of ammonia], *Succo Heliotropii* [juice of *Chrozophora tinctoria*], *Coccionella* [cochineal] or *Resina gallarum* [gall-apple resin], but with *Saccharo Saturni* [sugar of lead, lead acetate] became milk-coloured; thus it should be pure, but contain some lime, which is usual here.

Roma church was passed on the way to Visby.

Linden trees are rare here, even rarer are alders, and there are no beeches.

The farmers were hurrying on with their mowing, and most meadows had already been harvested. The corn-fields that were not altogether dried up had become green in patches after the rain. On both sides of the road one saw beautiful meadows and oak groves.

This night we rested in the town of Visby once more, thus we had travelled around the whole island after our departure on the 25th of June.

July 17 [p. 293]

When we had rested after our journey, we visited the beaches around Visby, where we gathered corals and petrifactions, but I will not specify these. The reader is referred to a *Disputatio de Coralliis Balthicus Ups. 1745* [see 2 July]; the same corals were found here as found before, except that here *Porpiter* were in quantity.

Carduus nutans Fl. Svec. 654 [*C. nutans* L.] grew in the streets; the heads were not compressed or rounded, but wide open. The leaves ran down the stem half-way to the leaf below but by no means the whole way, just as in the common thistle [*Cirsium vulgare* (Savi) Ten.].

The old houses in the town had pitch-black stone walls, which at first we thought was due to fire, smoke, soot or dust, but finally, when we looked at it with a microscope [i.e. a lens], we found out that this black colour was *Florae filia* rather than *temporis* [a daughter of Flora rather than of time], since it was a *Byssus pulverulenta atra* [*B. antiquitatis* L. = *Lepraria antiquitatis* (L.) Ach., but possibly not of organic origin; cf. also *Taxon, 16:* 185 (1967)], consisting of very small particles.

Marl was shown to us by an English sea captain, who said this was the real English marl. It was of two kinds, one taken at Gothum river, the other from Närsholmen; the former clay has already been described [see 2 July] and both were nothing but ordinary clay mixed with bog lime, of which only the latter seethed when water was added, in the same way as the true marl, but bog lime can probably never become a fertilizing marl, since the bog lime districts are the most sterile places in the whole world.

July 18 [p. 294]

We searched for corals and petrifactions today too, and we also visited the gardens in Visby, where the mulberry trees and the walnut trees had perished during the last cold winters.

The winter does not become cold here at the same time as in the rest of Sweden but continues more like a cloudy and wet autumn until springtime, when it gets cold.

Colica hypochondriaca is the most common endemic illness in the country. The symptoms are *flatulentia, tormina, alvi constipatio, ructus, pressio versus Jugulum, cervicem, humeros cum rigiditate semper, non vero semper cum anxietate* [flatulence, stomach-ache, constipation, belching, pressure on the throat, etc.].

There were many household remedies for this affliction, of which one was remarkable; a man, who had the disease for a very long time and had tried other preparations in vain, was advised to take *Decoctum foliorum Hepaticae in cerevisia secundaria* [decoction of the leaves of *Anemone hepatica* in beer] to drink morning and evening, furthermore to ingest *Oleum Terebinth* [oil of turpentine] morning and evening, starting with one drop and increasing to 7 drops *pro dosi,* then decreasing the dose for seven days till once more one drop *pro dosi* was taken. This remedy, however simple and strange it may seem, cured the patient.

July 19 [p. 295]

Today we went to church in Visby. The women were dressed in black and wore muffs when they partook of the Holy Communion, although the dog-days [the hottest time of the year, formerly associated with the rising of the Dog-star] were at their height.

July 20 [p. 295]

Vible is a farm with a little lime kiln, owned by Inspector [Carl] Lundmarck, and situated ½ a mile south of Visby. I went there for a few days, since we had to wait for a ship to cross to Sweden.

The limestone that is quarried at Vible is very beautiful and light grey, full of small grains shaped like little kidneys, white like quartz, and said to yield more lime than the rest of the stone. When the limestone was put into the oven, it was arranged as 8 pillars the better to allow the flames to play around it. On the first day of burning rather little fuel was used, so as not to make the heat too strong, and by the second day the limestone looked as if wet or turned into glass; if this had continued, all the limestone would have been changed into glass, and thus spoilt; accordingly, as soon as the workers notice the vitrification, they add more fuel to make the heat stronger and prevent this. It is rather amazing that limestone can be stopped from becoming glass through more intense heat, for the strongest effect of heat on bodies is to convert them into glass, and that limestone can be turned into glass like flint but not flint into lime; this is why limestone makes ore more fluid in the furnace, and why this flux must not be heated too much. I saw a third remarkable thing, but not as clearly as I had wished, and I would like others to investigate it. I saw that, on the road where in the morning the burnt lime had been taken to the water to be slaked, there were handfuls of slaked white lime; when I asked the reason for this, since slaked lime is not transported on this road, I received the answer that when small stones fall off the cartloads, they are slaked by the sun. There was no rain today, and it was very hot. The workers insisted that sunshine slakes burnt lime quite as well as water, however much I contradicted them. The lime that is burnt here is among the best in the whole country; it is wholly white, when touched with the fingers as soft as powder, without any sand in it, and can be combined with more sand when used as mortar than any other lime; mortar made from this lime can be used to bind bigger and heavier stones than any other.

We saw stalactites growing in a waterfall by a sawmill close to the seashore, where the limy water continually ran down a moss or *Conferva*, which was encrusted with lime, and this stone-crust was coated with small particles which grew on the rock and was still soft enough to be removed from the rock with a knife; between the grains were small holes, in which we found *Cancri pulices fluviatiles dicti* (*It. Oel. 42*) [see entry of June 2: *Cancer pulex* L. = *Gammarus pulex* (L.)].

Byssus [*B. lanuginosa flava* L., *Fl. Suec.* 390 no. 1129 (1746), 2nd ed. 390 no. 1129 (1755)], looking like velvet, rather tough and pale yellow, grew densely and firmly where the land dropped like a wall precipitously to the sea, down which water ran continuously here. This moss was encrusted with lime at the base.

Melica petalis exterioribus ciliatis, Fl. Svec. 56 [*M. ciliata* L.] is a grass that has not been known to grow in Sweden before and was found here between the piles of stone and the juniper bushes on the most sterile ground between Vible and the sawmill [see p. 11].

Carduus acaulis, calyce glabro, Fl. Svec. 656 [*C. acaulos* L. = *Cirsium acaule* Scop.] is a thistle that we have seen growing all over the island, but since we had nowhere seen its flowers until now, we had considered it to be a little *Carlina* or some known *Carduus*, but when we saw it here in bloom we realized that this is a species hitherto unknown in Sweden. *Folia oblonga, alternatim pinnatifida. Caulis duos pollices altus, et saepius fere nullus, terminatus capitulo seu calyce vix ventricoso, oblongo, levi, imbricato: squamis apice mucronatis, neque rigidis neque pungentibus: interioribus apice purpurascentibus. Corolla universalis purpurea: flosculis quinquefidis: laciniis duabus profundius divisis* [see p. 10].

Alisma fructa globoso undique echinato, Fl. Svec. 301 [*A. ranunculoides* L. = *Baldellia ranunculoides* (L.) Parl.] grew between the lime kiln and Vible farm in a ditch with a little stream; we have never seen it before except in France. It grew together with the common *Alisma* or water plantain [*A. plantago-aquatica* L.], but in deeper water; the umbel was simple, but the flowers were bigger and whiter. Just before noon we saw the stream in one place looking all white from the flowers, but at 7 p.m., when we returned to look for it once more in that place, all the flowers had disappeared, and we would not have found it if we had not recognized it by its leaves. It had a much more pungent smell than the ordinary water plantain, hence this made us suggest that its harmful effect on horses might be more pronounced, and Inspector [Carl] Lundmarck was asked to investigate this and report the results to the Royal Academy of Sciences.

Oniscus cauda obtusa integerrima [*O. armadillo* L.], a kind of woodlouse, was found in the woods; it was completely black except for the edges of the segments, which were white, and a white spot under the back legs on each side. The tail was obtuse and not cloven; when one touched the little animal, it coiled like an *Armadillo,* so that it looked more like a purgative pill than an insect [see p. 13].

The he-goats had a formation under the chin looking like teats, just as pigs sometimes have, which the people here take as a propitious sign.

The following plants are called iron-roots (järnrötter) by the field-workers here: *Ononis* [*O. spinosa* L.], *Centaurea calycibus ciliatis subrotundis foliis pinnatifidis* [*C. scabiosa* L.] and *Anchusa* [*A. officinalis* L.], all of which grow in the fields, but the roots of the first are the most troublesome.

"Kräkfötter" (crowfoot) was the local name for *Aster montanus luteus salicis glabro folio C.B.* [*Inula salicina* L.] that grew everywhere in the meadows, and was not loved by the farmers.

Besides these, the following plants grew here: *Helleborine flore purpureo* [*Epipactis atrorubens* (Hoffm.) Schultes] in the dry meadows, *Helleborine flore albo* [*E. palustris* (L.) Crantz] and *Orchis muscam referens* [*Ophrys insectifera* L.]. *Sedum rupestre repens foliis compressis, Fl. Svec. 388* [*S. rupestre* L.] grew on the rocks. *Centaurium minus* [*C. erythraea* Rafn], *Astragalus Riv.* [*A. glycyphyllos* L.] and "Ludd-Tatelen", *Fl. Svec. 67* [*Holcus lanatus* L.] were found here and there.

July 21 [p. 299]

The water in the well on Vible farm was so salt that it could hardly be used at all, which is quite remarkable, since the bottom of the well was many fathoms above the sea level. *Saccharum Saturni* [sugar of lead, lead acetate] made the water all white and opaque; the water in Visby is also whitened by this lead salt, but still remains fairly clear. When *solutio lunae* [silver nitrate solution] was added to the Vible water it turned white and coagulated partially, and the white coagulate was precipitated to the bottom.

Assessor [Johan] Pihl told us about the seal hunting on Sandö, which is done with square nets, as described earlier [see entry of June 27], and with a harpoon. The harpoon is a weapon with a wooden shaft that is as thick as the handle of an axe and about 1½ ells [3 feet, 90 cm] long. The head of the harpoon is as thick as a finger, 1½ quarter-ells [9 inches, 22 cm] long and has 2 barbs that are thin and sharp and 1½ inches [3.8 cm] long; its hollow end is screwed on to the shaft. A rope 9 fathoms [52 feet, 16 m] long is attached to the harpoon and wound round the shaft. When the seal is asleep on the shore, the farmers sneak up close from the leeward so that the seal will not be alarmed by their smell and make its escape immediately. When they are at a distance of about 20 steps from the seal, they run towards it throwing their harpoons while retaining a grip on the rope with their free hand; they let the seal make off into the sea on a slack line, they call their companions for help and, when the seal is tired, they draw it in to the shore and a few men go out and kill it. See entry of June 27 for illustration [see Plate 6].

Assessor Pihl also gave us a frog that was found in the sandstone at Bursvik, together with a complete report about it which can be seen in the *Handlingar* of the Royal Academy of Sciences for 1741, p. 248.

July 22 [p. 300]

We tried every day to leave Gotland, but had to stay here like prisoners, because no ship was ready to leave—and to trust one's life to the mail boat, which was small, frail, old and unsound, and, still worse, to its captain, would have been very adventurous. Nevertheless we had to make a deal with him, but I do not know what confused him, since he failed us and left without our knowing. If those who are responsible for this mail-transport would arrange matters in another way, they would do the inhabitants a great service.

Dianthus floribus aggregatis, squamis calycinis lanceolatis longitudine tubi, Fl. Lapp. 345 [*D. armeria* L.], was found among the crops in a field near Visby; it has also been sent to us from Ballenbo. The petals were small, lanceolate and had many small teeth at the margin. This was hitherto unknown in Sweden.

Cakile maritima angustiore folio Tournef. [*Raphanus* 569 *Kakile* of Index = *C. maritima* L.] and *Salsola* [*S.* 206 *vulgaris* of Index = *S. kali* L.] grew on the beach; I had found both before on the beaches of Skåne.

July 23 [p. 301]

A potter has recently opened a workshop here, since there has not been one in the country before. He obtains his clay from Roma monastery and it seems to be very good.

The brickworks which were established here recently have encountered difficulties in getting a good clay, for the lime-mixed clay produced bricks of which as much as a third cracked as soon as they got wet, since burnt lime absorbs damp, becomes hot and expands, causing the bricks to split.

Conferva Plinii [*C. rivularis* L. = *Cladophora rivularis* (L.) van den Hoek] grew here near Högklint, where the slope is steepest towards the sea and where the water runs continuously and was so long that I have never seen its like, namely more than 5 ells [10 feet, 3 m] long.

Pyrites ferri cubicus, beautiful and even, was found in the clay behind Högklint.

When the grain starts to rot, hazel rods are put into it as a remedy.

The meadows are cleared from hazel bushes in a very neat manner; they are cut off some ells above the ground; but if they are cut off at the root, they do not disappear.

Nettles are cut and collected in great amounts; they are dried for the winter, when they provide the sheep with tasty and wholesome fodder.

Visby's main trade is in beams, boards, lime, beautiful white tar, wool and the tasty Gotland mutton. Several merchants spend all the summer in the countryside, near the harbours, trading with the farmers.

July 24 [p. 302]

Bishop [Georg] Wallin's beautiful library, very nicely situated in the middle of his garden, was visited today and his many antiquities and curiosities seen.

At last we found a ship; thus we said good-bye to the gentlemen of Visby, who had almost competed with each other in their cordiality towards us.

Night fell, and the sailors were not yet ready; thus all night we remained *antipodes in urbe* [a phrase used by Seneca, one of Linnaeus's favourite authors, for revellers who turned night into day].

July 25 [p. 302]

At half past five in the morning we went on board. At the risk of our lives we left the harbour in a furious sea, our friends and Visby disappeared, the Karlsö islands appeared, a northern wind started to whistle, the waves raged, the ship was thrown between the roaring billows, Gotland disappeared, my companions became sea-sick, the rigging burst, our hearts filled with despair and we committed our fate to God's hands. . . .

[*Linnaeus inserted out of chronological sequence his narrative for July 25-27; this deals with the landing of the travellers at Böda in north Öland on July 25, their journey southward from Horn to Färjestad on July 26 and their crossing to Kalmar on July 27; see pp. 105-107 above.*]

The Journey from Gotland

SMÅLAND

Kalmar town did not keep us for long on our return from Öland. I left the same day, in the afternoon (July 27).

The stone houses, which were made of Öland stone instead of brick, e.g. the Laboratorium, the Residence, Borgholm Castle, have the disadvantage of being damp, which destroys the wallpaper in spite of the fact that the walls are daubed and whitened and a hundred years old. This is the reason why the Öland farmers build their outhouses of stone, but their dwelling houses of timber. Some had tried to avoid having damp walls by covering the inner side of the stone wall with a layer of brick.

A supposedly infallible remedy for diarrhoea and dysentery was the following: Take three whole eggs, 1 nutmeg, 1 quintin [1¼ scruple, 1.7 g] mace, 1 quintin saffron, 4 spoonfuls of wheat flour. The ingredients are pulverized, mixed with the eggs, and red chalk is added to make a firm dough; the dough is shaped into little cakes which are baked on a hot brick in the oven. The cakes are then pulverized and ½ lod [i.e. ¼ oz, 7 g] of the powder is ingested in beer or, better, red wine for a dose.

Some people have thought very highly of the possibilities of forecasting the weather using this chart:

Hours	*Reigns*	*Results*
Morning 6-12	Fire	warm weather with southerly winds
Daytime 12-6	Air	windy weather with west wind
Evening 6-12	Water	wet weather with east wind
Night 12-6	Earth	cold weather with northerly wind

That is, when the moon changes in the morning, between 6 and 12 o'clock before mid-day or when Fire reigns, the following quarter will usually be warm, with south winds; when the quarterly change occurs in the day between 12 and 6 o'clock, when Air reigns, this gives strong westerly wind; if the change is in the evening between 6 and 12 o'clock, when Water rules, there will be rain and east winds, and if it occurs at night, between 12 and 6, or under the influence of Earth, northern winds will bring cold.

"Wau" or "Wouw", a beautiful dye-weed called *Reseda foliis simplicibus lanceolatis integerrimis Fl. Svec. 439* [*R. luteola* L.], was found here growing wild close to the town on the walls; I have seen it only in Skåne around Lund before; since the plant grows wild here there is no need to import it. It could be harvested and sent to the dye-works, where it is used for yellow colour.

Man's Blood, *It. Oel. 35* [see entry of May 29], *Fl. Svec. 251* [*Sambucus ebulus* L.], was in full flower now and pervaded the whole district with its scent.

Campanula foliis ovato-lanceolatis, caule simplicissimo, floribus secundis sparsis, Fl. Svec. 18 [*C.* 180 *maxima* of Index = *C. latifolia* L.] grew wild at Masbo church.

July 28 [p. 304]

After arriving at Kråkenäs, near Växjö, I had to rest, for I was exhausted from the daily toil of the journey during the last two months and I asked for nothing during the following few days.

Aug. 1 [p. 305]

Today I journeyed to Orraryd. There were innumerable barrows and upright stones along the road between Skye and Torsås.

Inglinge mound stands near Ingelstad, at a few gunshots' distance south of the Inn; it was about 4 fathoms [24 ft, 7.4 m] high and 150 paces in circumference. On top of it was an upright stone 3 ells [6 ft, 1.8 m] high and almost as broad. A big globular somewhat depressed stone (*depressus globosus*) 5 ells [10 ft, 3 m] in circumference, adorned with pictures but without letters, lay at the base of the stone on the south side. Along this mound there was another one, half as big, on the southern side, which had later been excavated. Both consisted of stones covered with turf. On the southern side of the larger one grew almost nothing but sheep-grass, *Fl. Svec. 95* [*Festuca ovina* L.].

Arnica, Fl. Svec. 684 [*Doronicum* 684 *Arnica* of Index = *Arnica montana* L.] grew in all the meadows and *Dortmanna Rudbeckii, Fl. Svec. 714* [*Lobelia dortmanna* L.] was common in the water.

The rye-fields had tall and beautiful rye but the farmers complained that the ears were "laddered" since the rain had prevented the rye giving off its dust. Thus they understood *practica sexum plantarum* without any theoretical instruction. They call the ears "laddered" because many grains were missing, thus giving the ear the appearance of a ladder; this came about because the continuous rain when the rye was in flower had prevented the pollen falling on the stigmas of the females in the spikes, and hence no seeds had developed.

Orraryd gave us night lodgings.

August 2 [p. 306]

The garden paths in Orraryd were covered with sawdust instead of sand, which made the bushes of box thrive very well.

A woman had been stung by an adder; I prescribed a large quantity of olive oil for her to be taken often and in large quantities but a few days later I heard that this remedy, much praised by the English, had no effect whatsoever, although the woman was bitten by a grey snake (*Vipera*) [*Vipera berus* L.] only 6 hours prior to treatment. Thus physicians and apothecaries should rather supply our pharmacies with *radicem senegae* [root of *Polygala senega* L.] from Virginia.

Asps, a kind of red snake [a reddish variant of *Vipera berus* L.] said to be short and thick but very quick and with a deadly bite, were often seen here. Whoever sees one of these animals will do well to describe this undescribed animal; in particular the number of scales between the chin and the tail on the underside should be specified.

The slow-worm [*Anguis fragilis* L.] is an interesting snake which breaks when it is hit. The belly is black, the sides purple, the back grey with a dark line separating the back from the sides. The people believe that this snake stings only at noon, but the teeth show that it is quite harmless.

August 3 [p. 307]

Aphides serratulae foliis dentatis spinosis [*Aphis cardui* L. = *Brachycaudus cardui* (L.), for which *Compositae* are secondary hosts] sat in great numbers on the field thistles [*Serratula arvensis* L. = *Cirsium arvense* (L.) Scop.]. These animals were dark, but with legs and antennae white with black tips, the back antennae entirely black; on these *Aphides* was a little *Ichneumon Aphidum*, not larger than a minute mosquito, all black except for legs and belly which were pale at the base. It flexes its belly under the breast like a plough and injects its eggs *in ano* or in the back horns of the little *Aphid*; it performed its task with much devotion and did not interrupt its work although we broke off the branch it was sitting on and shook this.

Black soil like the black dye mentioned above [see entry of May 27] had been sent to me from Lenhovda, but it was only a black mud containing iron ochre collected at Borgsjön in Flisby; it is used for black colour after mordanting the yarn with a decoction of the inner layers of alder bark.

Jacea nigra pratensis latifolia C.B. Fl. Svec. 709 [*Centaurea* 709 *vulgaris* of Index = *C. jacea* L.] had been tried in the dye-works in Växjö for yellow instead of sawwort; it dyes yellow but less perfectly than *Serratula* [*S. tinctoria* L.].

August 4 [p. 307]

Araby, a farm owned by the District Governor [Anders] Unge, situated half a mile from Växjö, was visited today for its ancient mounds and stones and the Helgasjö lake.

Some people believed that Helgö here was the *Helicon* of the ancients, upon which stood Wotan's temple. The building could not have been very impressive, however, if those walls and cellars which are still discernible were its ruins.

Milsöar is the name of two little islands lying at the side and thought to be the *Insulae musarum veterum*.

Leva spring, situated across the lake west of Araby, at the second milestone on the right side of the road to Ör, has been a stone-built sacred spring once. This was where pilgrims performed their ablutions before coming to Wotan's temple, and *Silviae* town is supposed to have been situated here.

A small high island, flattened on top, which I leave to the antiquarians together with the above, stands not far from Araby.

August 5 [p. 308]

Superstition occurs in all parts of the world but not everywhere equally or the same; and it is usually commoner in provinces remote from the capitals, which are rarely visited by strangers. I will mention here some popular beliefs which I have heard in Småland, Kalmar and along the border to Skåne, leaving out those I have heard by the way in other provinces, although many superstitious beliefs are common to Småland and places far away, such as Dalarne and Västerbotten.

The bride should try to recognize the bridegroom at the wedding before he sees her; for then she will hold sway over him.

The bride brings some food in her pocket, which she dispenses to the poor; for each alms she gives, she will avoid one future calamity (hence she gives her alms with much charity), which will instead befall the receiver (who will therefore often dismiss the gift), who already is unlucky enough.

When the bride and bridegroom ride to church, he holds her reins, so that no one shall come between them to prevent him from keeping her for himself.

The bride does not tie her shoes before the wedding, so that she can give birth as easily as she slips off her shoes.

If the bride has a coin in her shoes during the wedding, she will never lack money in the future.

At the wedding, the bride will try to put her feet a little in front of the bridegroom's, since this will give her power over him.

The bride touches a bare spot of her body, as soon as she comes to the bridal chair after the wedding ceremony, with as many fingers as the number of children she desires.

The bride on coming home from the church goes immediately with the bridegroom into barns and stables, so that their cattle and horses will thrive.

The bride tastes at once all food which is prepared, to keep her from becoming greedy.

Bride and bridegroom eat from the same plate to ensure agreement in the future.

As soon as the bride has returned from church she puts her head rapidly into a *fecundina equae* in order to have easy and fortunate childbirths.

The bride tries to keep awake longer than the bridegroom so that she will not die before him.

If the fire crackles and bangs very much, this means that someone will die.

When someone shivers violently, people believe Death walks over the grave.

If children or dogs dig outside the house, this means that someone will die soon.

If a grave falls in within the graveyard, this means that someone of the same kin will die soon.

When the owls call outside the houses, death or fire will ensue.

If a horse or bell-cow dies, this means that father or mother will die soon.

If the corpse leans over to the right in the coffin, this means that a male kinsman will join him next; if to the left, the next one to die will be a woman.

In some places a candle is put into the hands of the dead (*reliquae papismi*) [relics of popery].

The old man's dearest possession, e.g. a pipe, a tobacco pouch, money, tinderbox, is often put into the coffin to prevent him from haunting the place.

On Christmas night two candles are left burning the whole night upon the table; if one or other goes out, this means that father or mother will die.

Straw should always be left on the floor during Christmas, and crosses made of this straw are laid outside each door, as well as the tables, so that all will be blessed. Crosses of this straw are also thrown on the fields, and fruit trees are encircled with it, so that they will be blessed with much fruit. If grains or seeds are found under the table on Christmas morning, this is supposed to be a sign of a fruitful year to come (not of careless threshing).

On Midsummer's Day no one picks anything green from the earth, not even to smell a flower, in order not to be infected by "carrion worms" which swarm at that time.

Berries growing at the church road, where corpses are carried, must not be eaten, otherwise one might be infected by "carrion worms". (Can one eat berries from roads where crayfish have been carried and not have cancer, which is the same disease as "carrion worm"?)

No one spins on Thursday evenings, since the spinning would continue all night.

On Maundy Thursday crosses are put on all doors, to prevent witches from doing harm. If one shoots in the air, all the crones on their way to Blåkulla fall down when they hear the sound of the gun.

And infinitely much more of similar rubbish. To eradicate these things there is no better way than to instruct future theologians well in natural history and natural philosophy, since nothing serves so well to check superstitions as the contempt of the clergy. It is truly remarkable how these and many other beliefs have been retained by our nation from times immemorial. Part of them one finds among poets soon after or before Christ's time; a part is a relic from heathendom in Sweden; a part from popish times; a part invented later. It would be a very nice endeavour if someone would collect a considerable amount of these popular beliefs and trace where and when they first arose.

August 6 [p. 312]

Today we travelled to Åby; we did not see anything remarkable on the way, only a few man-made cairns here and there in the woods as a testimony that here, where now grow vast forests of spruce and fir, were once fields and cultivated land; this is often seen in woods in Småland, from which one can possibly conclude that the land has been formerly more cultivated and densely populated, but I do not know what kind of plague has eradicated so many people, perhaps the Black Death; it is known from history that in the year 1315 one third of the population succumbed, and in 1345 half of the people died, and in 1350 still more.

August 7 [p. 312]

This day we spent in Växjö.

The wise woman Ingeborg of the parishes of Mjärhult and Virestad was consulted from all over the country, like an oracle, and had greater fame in medicine than many a doctor who has done nothing but study and practice; we did our best to probe her wisdom from listening to her stories and we arrived at some understanding of her theory and practice. She believed that Lucifer and his followers had been expelled from heaven unto the earth. Some of the followers had their abode in the water under the name of "Näcker" [water sprites], some under houses with the name of "Tomtegubbar", some in reeds and among trees called "Elwar" [elves], in woods as "Skagsnufwor" or "Rå" [wood sprites]. She believed that every human has a wraith, which follows him like a shadow, and this wraith walked perpendicularly under the earth, just as man walks on the earth, and the feet of the wraith were always attached to the man's feet, like the reflections of animals, trees and mountains in calm water. She also believed that a man and his wraith are connected to each other so that, when the earthly man suffers, the subterranean one does so too, and *vice versa*; when the subterranean being was harmed, the man on earth suffers (*nam quod est superius est sicut inferius*); she believed that when men walk and their *Antipodes* [their underground wraiths] happen to pass the home of a goblin, or an elf, or some other sprite, the subterranean man is hurt and the man on earth will be diseased. But as far as her *Semiotica* or knowledge of diseases and their causes was concerned, it surpassed mine widely, as well as that of other medical men; when someone was ill, she did not need to see the patient, nor did she have to ask about his constitution, his temperament, pulse, symptoms, or diet; but it sufficed for her to see a stocking, a garter or any garment that the sick person had worn, and from this she would draw conclusions as to the cause of the disease. Her *Pathologie* or verdict was most often that the patient had passed water or slept in some locality dedicated to a spirit, or taken something from a holy tree, or contracted the disease from air, water, fire or earth, which is a wonderful theory for anyone who wants to live well and cares about his health. But the cure was very different from those taught by the Medical Faculty; when the cause is immaterial, the cure has to be immaterial too: for instance the patient should leave home mute and fasting on three mornings, or three Thursday nights, mostly walking northwards or to some stream flowing northwards, or to some tree, and then make apologies and give offerings of milk or something similar. When this has to be practised on certain weekdays only, so that the cure will take three weeks, I think the correct Latin expression is to cure *per expectationem.* She had often been in trouble because of her practice, with judges as well as with the clergy, since they believed she was practising witchcraft; this she would not confess to, defending herself with the words of Christ to the Pharisees when they accused him of casting out the devils through Belzebub [the prince of the devils], saying—"Show me a single man to whom I have caused some evil, and I will confess to everything, but if I have done some good, how could I do anything good through the finger of the Evil One? For then his kingdom would be divided against itself." Otherwise she led a devout life, was a regular church attendant and was kind and respectful towards all. Until her recent death she always had a lot of visitors from all parts of the country, for the farmers believed in her like in an oracle.

August 8 [p. 314]

Today we travelled from Växjö to Stenbrohult, which is almost on the border of Skåne.

Husaby, 2 miles from Växjö, had an iron works of considerable size with a blast furnace, foundry, hammers and mills.

The iron ore was obtained from the bottom of Åsnen, a little lake nearby, and looked like big iron hailstones, which were soft inside almost like a *Geodes*; this ore was extracted during the winter from the ice, in the summer from a raft using an iron rake with which the ore was scraped from the bottom into a sieve; after which it was used without previous roasting. The flux was a dark spar- and mica-containing greystone which was roasted and had a little iron. For each production, 20 barrels (20 skeps) of ore, 2 barrels of flux and 20 barrels of charcoal were used, 10 times a day. The iron is cold-short [brittle when cold], most suitable for cannons, cauldrons, stoves, ploughshares, etc. Iron stoves are bought in great quantity by the people of Skåne.

"Tagelmyra" lay on both sides of the road, three miles from Växjö; here grew *Betula nana Fl. Svec. 777* [*B. nana* L.] ; the peasants use the twigs for whisks to scour the milk-pans in summer, when the milk sours quickly.

The woods the whole way were of fir, spruce, juniper, heather, whortle-berries, bearberries and crowberries, all green both winter and summer. The meadows were very pleasant with beautiful deciduous trees, especially since the farmers would never tolerate fir, spruce or juniper in them.

August 9 [p. 315]

Stenbrohult church stands on the shore of the big lake Möckeln, which here makes a large bay and produces one of the pleasantest of sites. The tall alder trees prevent the lake from encroaching on the land, as they grow at the water's edge.

The garden which my father, the clergyman Nils Linnaeus, laid out here had more kinds of plants than any other garden in Småland, and this garden gave me with my mother's milk a burning love for plants.

Corispermum foliis alternis Hort. Cliff. or *Corispermum hyssopifolium Juss. act. paris. 1712 p.244.t.10* [*C. hyssopifolium* L.] was then in flower in the garden and showed us something strange; namely that the nethermost flowers had 4 stamens, the middle ones 3, then 2 and the other flowers had only one stamen.

There were many plants growing wild around Stenbrohult which otherwise are quite uncommon in Sweden. I had joy in finding them in the same places as where I grew up with them; the rarest were the following: *Dortmanna Rudb. Fl. Svec. 214* [*Lobelia dortmanna* L.], *Plantago scapo unifloro Fl. Svec. 128* [*P.* 128 *monanthos* of Index = *P. uniflora* L. = *Littorella uniflora* (L.) Asch.], *Calamistrum Fl. 996* [*Marsilea* 996 *calamaria* of Index = *Isoetes lacustris* L.], *Spongia Fl. 1132* [*Spongia lacustris* L.], *Sparganium natans Fl. Svec. 771* [*S. natans* L.], which all grew in the lake, not far from the shore. *Gramen ossifragum Fl. Svec. 268* [*Anthericum ossifragum* L. = *Narthecium ossifragum* (L.) Hudson] grew between the outhouses, *Hydrocotyle* [*H. vulgaris* L.] in the south pasture, *Radiola Fl. Svec. 256* [*Linum radiola* L. = *Radiola linoides* Roth] at the footpath, *Scheuchzeria* [*S. palustris* L.] on the marsh-meadow, *Elatine* [*E. hydropiper* L.] on the shores, *Jungermannia hypophyllum Fl. 930* [*J. epiphylla* L. = *Pellia epiphylla* (L.) Lindb.] at the mill-stream with hitherto unnoticed stamens, which were now easily discernible. Hornbeam or *Carpinus* [*C. betulus* L.] grew in the south enclosure, *Arnica* [*A. montana* L.] at the carpenter's workshop, *Aphanes* [*A. arvensis* L.] in the hopgarden, *Riccia* [*R. sorocarpa* Bisch. ?] under the beeches and *Lycopodium palustre repens clava singulum Fl. 862* [*L.* 862 *clava foliosa* of Index = *L. inundatum* L.] on the shores of Gåsön together with *Pneumonanthe Fl. 202* [*Gentiana pneumonanthe* L.].

August 10 [p. 317]

The people in this district, as in most Småland parishes on the border of Skåne and Blekinge, such as Virestad, Torsås and Urshult, were in both sexes generally bigger than in other places, which is probably due to the old Gothic unmixed stock, since strangers are seldom seen and a farmer rarely gives his daughter in marriage to someone who is not born in the parish.

The farmers wear a special costume here. The menfolk were clad in long black "Walmarströjor" [a kind of jacket] with a brown cloth edging at each seam. The shirt had a large collar, which lay outside the jacket, with lace-like decorations on the collar edges and the cuffs. The shoes were cut low and had heavy soles of birch bark and narrow heels, and were soled beneath with iron rivets. The women had covered their heads with a white neckerchief tied in the nape of the neck and shining from being polished with glass-stone. But the young girls had glittering ribbons in their hair, which is left uncovered; the breast is covered with a vest, sewn to the shirt and decorated with silver rings on the bosom, and under this is a blouse with wide sleeves. A red belt is worn slightly below the waist; it is tied and hangs down on the right side; the shoes were like the men's but cut lower on the sides. They keep the habits and garments of their forefathers in great esteem, and are as particular about them as any young lady about the modes from Paris. We learned how the farmers treat their children for scalp rashes. They wash them in cold water and then the children get fits, as has been described in *Vet. Akad. Handl.* 1742 p. 279.

August 11 [p. 318]

Some of my companions, J. Moraeus, H. J. Gahn & G. Dubois, who left me in Kalmar to travel to Skåne through Blekinge, returned now. They brought with them a rare plant, *Dainthus caule unifloro, foliis linearibus Fl. Svec. 343* [*D.* 343 *uniflorus* of Index = *D. arenarius* L.], not discovered in Sweden before; the leaves were linear, short, crowded together like in a brush; the stalk was little over an inch high, and had a single white fragrant much-cut flower. They had found it growing in the sand between Sölvesborg and Kristianstad, at the border between Skåne and Blekinge at Hedenryd Inn. They had also found wild barley, *Fl. Svec. 107* [*Hordeum murinum* L.], in large amounts on the walls near Landskrona.

Household remedies in the district were the following: swinestone [stink-stone, a kind of malodorous limestone] was used for dysentery, and was also remarkably good for stitch and ague; for *menstruis suppressis* the women put their head under the cloth which covers the newly brewed beer when it is fermenting. The strange cure is sometimes tried by the men too to prevent them from getting drunk during the rest of the day, but it takes a strong man to undergo that.

Frangula Fl. Svec. 194 [*Rhamnus frangula* L. = *Frangula alnus* Miller], here called "Törste", was used to make gunpowder; the peeled stalks are cut into rods of equal length, dried, laid *s.s.s.* [*stratum supra stratum,* layer upon layer] and burned; the fire is quenched with a wet cloth and the charcoal is said to be better than anything else for gunpowder.

I was eager to see Råshult farm before I left, because it was the first place I saw in this world.

Afterwards we visited Sake-sone-backe, well known for its golden sand. We saw clearly that this golden sand consists of mica from a very friable schistose stone, which glitters like gold; there was little left of this golden sand since all travellers take their portion, but it exists in sufficient quantity a quarter of a mile further away in Dihulte hage, east of the main road.

August 12 [p. 319]

We had intended to continue our journey today, but it rained the whole day as if the sky had opened, and we could not leave. The farmers were very worried since for 14 days they had been unable to cut their ripe corn and to harvest their scanty hay because of the rain.

The meadows, which here are often poor and full of moss, are said to improve best by enclosing them for two or three years and neither grazing nor harvesting them.

Farms and villages here in Småland are much more beautiful and charming than in many other districts, although they are not so big. Between the dry mountains and the vast spruce and fir forests one finds more lakes than in other places; the farmhouses lie in the middle of fields and meadows. The fields are yearly sown in the spring, without lying fallow, and only deciduous trees are allowed in the meadows, hence fir, spruce and juniper are always cleared out; the deciduous trees make the meadows look very pleasant, especially where oaks and beeches intermingle. They grow in little groves and are not much exposed to the winds which always disturbs those on the plains. The meadows are never grazed, hence there are many flowers in them; the farms are situated in the fields and not on the forest edge. The living room has a window on the southern side under the roof, which makes the room light.

Hares and hazel-hens [*Tetrastes bonasia* (L.)] are bigger here than in Norrland and even bigger than in Roslagen.

August 13 [p. 320]

Today we returned to Växjö.

Djö inn had a blast furnace in which ore from the lake was melted to yield iron; the ore came from Möckeln and other lakes.

Gålåsa inn was 2 miles from Stenbrohult; from here to Alvesta (Alvarstad) and the surrounding district stretches the Bråvalla moor, a somewhat raised flat ridge, famous for the battle during heathen times when the women of Virend defeated the Danes; there are still remnants from this time such as cairns, stone slabs, farms and place-names.

Poa panicula diffusa spiculis sex floris linearibus Fl. Svec. 73 [*P. aquatica* L. = *Glyceria maxima* (Hartm.) Holmberg], a very tall and large grass, perhaps the biggest among the Swedish grasses, grew wild at Husaby Bruk in the garden between dropwort and cowbane, below *Nepeta* [*N. cataria* L.] and *Marrubium* [*M. vulgare* L.]. The grass has hitherto remained undiscovered in Sweden; if it could be grown in a suitable soil it would be well worthwhile.

Pain in the breast is cured by some people in the following way: the patient is put into an oven and a magpie nest is burned therein, then the smoke cures the patient.

August 14 [p. 321]

Today we stayed in Växjö.

Assessor [Johan] Rothman told us about the water in lake Helga at the mouth of the mineral spring, for this water is excellent for bleaching brown cloth, and he doubted if there was better water for this purpose anywhere in Sweden, as he had had convincing proof of its value for many years.

"Hiertansfrögd" ["Heart's Delight"] is the local name for a *Mentha* which is often grown in the gardens; some of the burghers put it in a glass of water on the window-sill, where it grows and forms many roots and long shoots, so that the window is bathed in a lovely green, giving fragrance and shadow.

The products of the Kronoberg district are tar, pitch, potash, boards, beams, some iron, some corn, a little butter and also a little copper and gold.

The cheese, especially the one the clergymen make out of milk from the whole parish, is better than any other Swedish cheese and even than Dutch cheese.

August 15 [p. 322]

Today we rode to Kronoberg, half a mile from Växjö, where the District Governor resides.

Kronoberg castle, or rather the ruins of it, lies on a little island in Helga lake, not far from the Governor's residency. The castle was four-cornered; the old walls still stand with a round tower in each corner. The inner court of the castle was 37 paces broad and 50 paces long. The rooms inside are about 20 paces broad with several cellars, arches and cavities. The walls are 1½ fathoms [9 ft, 2.7 m] thick and 5-6 fathoms [30-36 ft, 9-11 m] high. They are made of greystone [a grey volcanic rock] outside but brick inside. In the northern gate under the arch hangs a lot of *Natrum nudum calcarium* like hoarfrost and on the sides were everywhere *stalactites calcis* like icicles with concentric layers. The castle was inhabited by rooks and crows only.

The district Governor, General-lieutenant [Anders] Koskull, told us about several economic matters worthy of notice; in three years he has persuaded the farmers to cultivate 4700 acres [about 1900 ha] of new fields in the country by ordering all farmers not cultivating fully to take in one "skieppeland" [area of land requiring 12 bushels of seed when sown] of new land each year.

Barley, he said, is sown in *Medium Martii* [mid March] in Östergötland, but here in Småland in *Medium Maji* [mid May]; notwithstanding, it ripens and is harvested at the same time in both places.

The right time for sowing should, according to him, be judged from the smell of the soil; as long as the soil is too wet in spring it has a raw and foul smell, but when the sun has prepared it and dried off the superfluous water, it no longer stinks and it is ready for sowing.

Sandy soil, which turns red after rain, he said, is the most sterile soil to cultivate and there is hardly any way of making it more fertile, which sounds probable, since all such soil holds iron; yellow and red colours in the mineral kingdom are mostly derived from iron.

Heather, he told us, is good for sheep, but the heather is trampled down by the sheep and is destroyed by their dung, which is very probable if the sheep are in too great abundance. In Skåne we have seen the heaths being burned every third year, in order to provide the sheep and the cattle with tender heather.

There was a spring at the residency in Kronoberg with excellent water for food, peas, tea, washing and dyeing.

Many large heaps of stone occurred everywhere here in the woods, although the soil was very shallow and the humus not much more than half a finger deep. Undoubtedly in this district, where formerly were fields, there was much more top-soil and this same top-soil nourished trees and herbs, which would inevitably be generated here again by the decomposition of trees and herbs if nothing else intervened; perhaps the burn-beating is the cause whereby the humus and the humus-forming plants have been destroyed?

August 16 [p. 324]

We spent Sunday in Växjö.

Juncus culmo nudo, foliis setaceis, capitilis [*capitulis*] *glomeratis aphyllis, Fl. Svec. 282* [*J.* 282 *turfosus* of Index = *J. squarrosus* L.] and *Erica, tetralix dicta, Fl. Svec. 310* [*E. tetralix* L.], both quite rare, grew at the camp.

August 17 [p. 324]

Today we left Växjö for Stockholm.

Silva spring lay at the main road, at the second milestone from Växjö on the right-hand side; it had been a well-walled sacred spring.

Gona forest, which is a beech forest mingling with Husaby forest, was traversed.

A spruce of strange form stood by the roadside ¾ of a mile from Matkulla inn on the right; the branches were much thicker than is usual for a spruce and had no little side twigs, but hung down like rods and gave it a peculiar appearance; the leaves or needles were not arranged like a comb, as in our ordinary spruce, but were irregularly dispersed along all sides. I have seen such a tree only once before, 12 July 1734, in the north of Älvdalen at Rotbro in Dalarna, a single tree. Can this be a *species hybrida* or blending of fir and spruce?

The flies were kept out cleverly in Starhult; a leafy twig was hung from the ceiling each evening, on which the flies would settle for the night and were carried out each morning.

Beer made from juniper is often drunk in Sweden, but here most of all, and made without much art; cold water is poured on to crushed juniper berries and left to draw, and the drink is soon ready without fermentation or boiling. This drink is good and pleasant as long as it is fresh but it must be used within eight days, and thus the quantities to be made should be adjusted to the rate of drinking. It purifies the blood like a *decoctum lignorum* and serves both for a wholesome drink and for medicine.

A beech grew at Boo inn and was the last we saw in Småland; further north we did not see beech except for a few at Omberg in Östergötland.

Leontodon Fl. Svec. 629 [*L. autumnalis* L.] was now commonly in flower; it is a true autumn flower which is never seen before midsummer.

We stayed the night in Vrekstad.

August 18 [p. 326]

The peasants had now begun the harvest. Four rows of sheaves with 6 sheaves in each row were set up on the stubble with the ears facing the sun. When they have stood like this for a week, they are piled up into stacks with the ears inward and the topmost sheaf at an angle, and left until carted away.

Tripes, or Mr de Geer's *Physapi elytris glaucis* [*Thrips physapus* L.] sat in thousands on the flowers, throwing their tails upwards and jumping with their feet.

Senecio foliis pinnatifidis denticulatis, florum radiis revolutis linearibus. Fl. Svec. 689 or *Jacobaea senecionis folio incano, perennis. Hall. jen. 177 t.3* [*Senecio* 689 *halleri* of Index = *S. sylvaticus* L.] grew on an old clearing in the wood ¼ of a mile from Vrekstad, and has not been seen in Sweden before. It was similar to the common *Senecio Fl. Svec. 690* [*S. jacobaea* L.] in height, branches, leaves, flowers and size—but the flowers had a backwards-rolled collar of cloven rays. This grew here in great abundance but one found no specimen without a *lepra* under the leaves, which consisted of yellow blisters which on bursting delivered an orange powder in which the orange grub of a fly could be found.

A slow-worm [*Anguis fragilis* L.] was seen at the road, similar to the one we saw on 2 August but with a dark spot on the head. The farm-girls sometimes stroke the back of this snake three times with their hand, on the supposition that the cattle will fare well when they are patted with a hand which has been made propitious in this way. The ancient *rei rusticae autores* knew the trick but do not mention the previous preparation of the hand with snakes.

Asilus corpore flavo hirsuto, pedibus nigris, alis glaucis [*A. flavus* L. = *Laphria flava* (L.)]. This big fly we found on the road; it had a curved snout with *antennis clavatis.* The tail was strangely shaped like a goat's foot [hence that of a male] with two pairs of spines, one pair curved; these spines surrounded a larger one, curved backwards and covered with a downy scale in the back.

The sandy road between Svenerum and Barnarp caused us some trouble.

A runic stone shaped like a half cone stood on our left hand along the road half a mile from Svenerums inn. One could only read

ᛁᛋᚱᚢᚾᚾᚢᛏᛅᛁ ᛒᛅ ᛋᛅ:ᛋᛏᛂᛁ ᚾ:ᛅᚠᛏᛅᚱ
ᛏ ᚢᚢᚱ ᛋᛁᚾ:ᚴᛁ·ᛘᛅᛁ·ᚱᛁᚢᚾ:ᚢᛅᚴᛋᛁᚾ:ᛒᛅᛚ

Linnaeus: *isrunnutaibesa : stein : efter*
tuur sin : ki . meiriun : uegsin : bel

Sm 75: *sikuarþr . let reisa : stein : eftiʀ . ————þur : sin : at menr———— .*
uek sin : be-an : kuþ hialbi ont : hans

"Sigvard had this stone raised in memory of . . . his father (or : brother).
The kinsmen . . . God may help his spirit."

This stone was cut out of greystone holding spar and mica and the spar had grown into the letters, and made them rough and illegible, a clear proof that stones do grow, although *per appositionem externam.*

Two stone slabs were erected close to each other along the road between Svenerum and Stiamo just before one comes to Getamo.

We rested in Barnarp during the night.

August 19 [p. 328]

Early in the morning we left Barnarp on a sandy road to Taberg.

Taberg is an iron mine above ground and could well be called "Småland's wonder" because it is unique in Sweden. The mountain is very high, to judge by eye 300-400 ells [600-800 ft, 180-240 m] high, and so steep on all sides that it can only with difficulty be climbed on the south and north side but not at all on the east side. This above-ground iron-mine has no *matrix,* but the whole mountain consists of pure ore, and the only thing to do in order to mine it is to blast away stones and let them fall down; those who do the blasting are paid only 2 *styver* for each shot. At the base of the mountain is a river of considerable breadth, along which blast furnaces are constructed; they have the best facilities for receiving ore. As we crawled up the mountain, we found some stone chips which may have been matrix for the ore and appeared to be a fine sandstone; but on top of the mountain we could see how the sand has fused to form a mineral which could not be told from the rest of the Taberg ore. From the top of Taberg we had a most beautiful view, for the mountain lifted up its head as a fir tree among juniper bushes. It is hard to conceive of any other origin for the ore than a sandy soil saturated with vitriol water, which has turned the sand into an ore, and when this land was covered by the roaring waves, the sea must have cut away the loose sand and left the firmer parts. In descending we saw on the north side *Chrysosplenium* [*Ch. alternifolium* L.], *Ophrys foliis cordatis* [*O. cordata* L. = *Listera cordata* (L.) R. Br.] and *Clavaria* [*C. pistillaris* Fries = *Clavariadelphus pistillaris* (Fries) Donk], a fungus which was almost flesh-coloured, slightly compressed, hollow and shaped like the lower part of a pen.

Charcoal-making stacks were seen in the woods aroung Taberg, made in the following way: a log, 4-5 ells [8-10 ft, 2.4-3 m] long, is dug down into the earth, 1 ell [2 ft, 60 cm] of it deep in the earth, the other part upright in the air. Around this "heart-log" the charring wood, which consists of logs 3-4 ells [6-8 ft, 1.8-2.4 m] long, is piled with the shorter logs towards the top and the smallest one on the summit, but a little hole is left around the heart log at the top. The whole stack is subsequently covered with spruce twigs, earth and turf, which are closely packed; the fire is introduced through the top hole, then the hole is closed and ventilation holes for the fire are made with an iron pole at the lower end of the stack, 1 ell [2 ft, 60 cm] above ground level, so that the smoke emerges from the stack out of these holes, but nowhere else. When the stack has burned for a fortnight and the smoke diminished, the stack is pulled down and the fire quenched to prevent the charcoal from burning to ashes in the open air. One charring stack will yield 60-80 *ryss* of charcoal, with 60 skepfuls for each *ryss.*

From Taberg we went back to Barnarp and then on to Jönköping.

Jönköping town was rather well built, with wooden houses; most houses are 2½ storeys high, of which the second and next storey overhang the lower one. The streets are broad. The town is well situated. On the north side is Lake Vättern, on the south side Lakes Lillsjön and Råksjön. To the west was the ruined castle with its walls still remaining. In the east were beautiful oak groves.

The Göta Court of Appeal stands in front of the market-place; we saw a big collection of witches' paraphernalia, such as books of black magic, which we read and found full of frivolities, of old and false prescriptions, idolatry, superstitious prayers and invocations to the devil, most of them in rhyme. Here we saw ties and knots of threads, silk, horsehair and roots. We blew the sacred horn without conjuring up the devil, and we milked the milk sticks without getting any milk. Here we saw the *Troll-tyre,* which was not manufactured by hags or devils, but is the third stomach of a ruminating animal. Here we saw the eagle's feet with outstretched claws, used by witches to scratch the belly of people suffering from colic; I see no reason to burn them for this practice any more than the Chinese, who puncture the abdomen, or the physicians, who for intense colic burn *moxa* on people's bellies. Then we saw the real witch instruments: knives, hammers and clubs that people had used for killing their enemies. We also saw *Hypersarcosi* from the stems of birch, beech or oak; these were pressed between two plates to make an impression of the true coin. It is hard to believe that this vegetable substance should keep its shape when exposed to melted tin or lead.

August 20 [p. 331]

Today we travelled from Jönköping to Gränna.

"Strandråg" (*Secale spiculis geminatis Fl. Svec. 106*) [*S.* 106 *maritimum* of Index = *Elymus arenarius* L.] grew outside the eastern toll gate in sand, creeping with its roots like couch-grass [*Agropyron repens* (L.) Beauv.]; probably the roots could be used for bread, like those of the couch-grass, in times of famine.

Winter rye had been sown recently in a few places. It was the custom to test the seed grain by putting a few grains between two peat-sods or two woollen cloths, which were put in the sun, often dampened; then after eight days one counted how much had germinated of the new or the old rye and from this one judged which seed was the most suitable for sowing

The gunpowder factory in Huskvarna was situated half a mile away from the town not far from the road on the right. It was well built and the craftsmen's houses and the smithies were built along streets. We looked over the place quickly. The stream descended from the high rocks above the gunpowder mills in a waterfall of considerable height, but it was not used to its full power.

We spent the night in Gränna, 4 miles from Jönköping.

August 21 [p. 331]

Early in the morning we sailed from Gränna to Visingsö.

The Vättern water is quite clear, so that one can see the bottom of the lake down to a considerable depth; this is clean and without any plants, hence it is rather poor in fish.

Visingsö lies a good half mile from the eastern shore or Gränna. On the west side across the lake one sees Västergötland, which lies not much further from the island than Småland on the east side. Visingsö is oblong, with a tall fir forest in the centre, much oak in the west, spruce in the south and birch in the east. The land was flat but raised in the middle, consisting of sand, mould and clay, and very fertile, which is here almost the only consolation for the farmers. Here we saw many fallow deer, because the country is the king's hunting-ground. The whole of this island belonged to the Counts of the Brahe family. The Brahe castle, which is now in ruins, stands on the east side almost halfway along the island. On the way into the courtyard were several niches; out of each one looked a bust of some Brahe who formerly owned this property. The castle was situated directly opposite Gränna, above which the Brahe family had built two stone houses on top of the rocks, one to the north-east, the other to the south-east from the castle, called Vadstena and Brahe House, which could answer a cannon salute from Brahe Castle. The church stood close to the castle and was very well adorned in all aspects by the Brahe family. There were several sculptures of the Brahe family; in the tombs stood quite a number of tin coffins; here rested Countess Sparre, upon whose coffin stood a four-sided casket containing the head of her husband, who had been beheaded during the reign of King Charles IX [1604-11]. On a tomb was a life-size sculpture of Count Pehr Brahe and his wife. In the vestry one saw a portrait of St. Britta [Saint Bridget or Birgetta of Sweden, d. 1373].

The school was situated further north beyond the wood. It was quite small and had formerly been a church, of which the old vestry was now the library, also presented by the Brahe family; upon it was a tower, topped by a four-cornered place like an observatory, from which one had a splendid view.

There was a *Gymnasium* nearby, and not far from this the vicarage.

A low stone stood a little to the south of the school, upon this was engraved:

PETRUS
BRAHE COMES
mitt på Landet
Anno 1673

A quite large stone, about 11 fathoms [66 ft, 20 m] in circumference, lay at a little distance from the vicarage facing the castle, and had the following inscription engraved, although the stone was hard greystone or *saxum*

PETRUS BRAHE Senator, primus comes in Wissingsborg, R.S. Drotzet: conjugem habuit. D.Beatam Stenbock sororem Reginae Catharinae Gustavi I Regis, feliciter rexit comitatum 29 annos, natus 1520 obiit 1591 venerandus senex.

Cui successit Filius D. ERICUS BRAHE C. i W.R.S. Senator, conjugem hab: Ducissam Elisabetham Brunsvicensem, rex:II annos.

Frater D. Comes MAGNUS BRAHE R.S. Drotztus II. Uxor. hab:comitiss: Brigitta D.e. Rasborg et D.Helena Bielke, praefuit comitat: 30 annos. hab: Fratres D. Ioh. & Ioachimum. D.Gustavus campimareschal &c. D.*Abraham Brahe* R.S.Senator ux: D.Elsa Gyllenstierna.

hab: Filium D.PETRUM BRAHE juniorem, R.S.Drotzet, et magni ducatus Finland. general: Gubernator. Qui anno 1655 comitat. successit.

et huc usque juvante Deo feliciter per annos XX comitat. administravit. Conjux D.Christina Catharina Stenbock, pia mater et benefactrix obiit anno 1650.

Iohan Werner Pictor 1665

In the fir wood along the road to the western shore of the lake south of the memorial the following words were cut into the fifth stone:

Magnus Brahe comes
natus 1564. Rexit annos 31. mortuus 1633.

There were many gardens on the island, and at the vicarage were walnut trees which had survived the last two hard winters.

The fields grew wheat, much rye and peas.

The wild plants were *Tragopogon* [*T. pratensis* L.], *Onopordum* [*O. acanthium* L.] and *Euphrasia odontites dicta* [*Odontites verna* (Bellardi) Dumort. subsp. *serotina* (Wettst.) E. F. Warb.].

Bidens corona seminum retrorsum aculeata, foliis lanceolatis amplexicaulibus, floribus nutantibus. Fl. Svec. 664 [*B.* 664 *integrifolia* of Index = *B. cernua* L.] grew at the fabulous Gibers hole, *cujus Folia omnia lanceolata, profunde serrata, simplicia, opposita, semiamplexicaulia, hinc fere perfoliata alis reflexis. Capitula nutantia serie squamarum calycis exteriore patenti-reflexa; flos radio carens; Semina quadridentata, retrorsum hispida retroflexis aculeis.* This kind seems to yield a better yellow dye than the common bur-marigold [*Bidens tripartita* L.].

Grasses that are suitable for sheep I have carefully noted during our travels, but the farmers here could tell us about a grass harmful to sheep, which grows in marshes and other places, especially during wet years, and kills the sheep when they eat it; some of the farmers said that it had yellow flowers and others described them as blue, and we were thus very eager to see it. At last a plant was shown to us, and we found that some of the flowers were yellow and others blue in the same plant. This is *Myosotis foliorum apicibus callosis Fl. Svec. 149, β foliis glabris* [*M.* 149 *arvensis aquatica* of Index = *M. scorpioides* L.], or "förgät mig ej" [forget-me-not]. The plant grows commonly in dry places, where it has small flowers and downy leaves and a thin and short stalk, and it is not known to be harmful to the sheep; but when this grows in wet places it becomes far bigger, with large flowers and glabrous leaves, like the specimen we examined; maybe this gives it a sharp taste, like most water plants, and makes it more dangerous for sheep.

There was no heather on the island.

Crows stay here the whole winter, and thus indicate the island's mild climate.

The deer were kept off the fields in a neat way with a linen thread hung on the top of the fence-poles, about ¼-ell [6 in, 15 cm] higher than the fence itself; the deer would not leap over this, for fear of a trap or an ambush.

The inhabitants kept many geese on the island.

There were stones here consisting of nothing but conglomerated silica gravel which had been coagulated with iron-containing water and was now a pure iron ore, wherein the iron could be seen as a steel margin along the edges.

Ancient tombs were strewn all over the higher parts of the island.

Apis nigra pedibus maxillisque flavis [*A. tumulorum* L.* = *Halictus tumulorum* (L.)] or a kind of small earth bee, flew over the tombs in great numbers in the hot sunshine and had their nests therein; they were very small, a quarter the size of a fly [*Musca domestica* L.?], completely black except the pale yellow feet and jaws, and had black eyes and antennae.

We spent the night on Visingsö.

August 22 [p. 336]

Early in the morning we returned to Gränna.

Gränna town is very small and lies between the high mountains in the west and Vättern in the east.

* The island of Visingsö in the southern part of lake Vättern is the type-locality of *Apis tumulorum* L., cf. L., *Fauna Suec.* 301 no. 1000 (1746), *Syst. Nat.* 10th ed. 1: 574 no. 2 (1758).—*W.T.S.*

A runic stone lay half a mile from Gränna to the left side of the road at Uppgränna village upon which was quite clear and legible

ᛋᚢᛁᚾ:ᚱᛁᛋᚦᛁ:ᛋᛏᛁᛅᚾ:ᚦᛅᛋᛁ:ᛅᚠᛏᛁᚱ:ᚬᛋᛚᛅᚴ:
ᛏᚢᚴ:ᛅᚠᛏᛁᛦ·ᚴᚢᛏᛅ·ᛋᚢᚾ:ᚼᚬᚾᛋ:ᛅᚾ:ᚬᛋᛚᛅᚴᛋ
ᚢᛅᛋ:ᛒᚱᚢᚦᛁᚱ·ᛋᚢᛁᚾᛋ:

Linnaeus: *suin : rispi : stian : þesi : eftir : oslak :*
tuk : eftiʀ . kuta . sun . hons : en : oslaks
uas : bruþir . suins

Sm 122: *: suin : rispi : stina : þesi : eftiʀ : oslak :*
auk : eftiʀ : kuta : sun : hons : en : oslaks :
uas : bruþir : suins

"Sven raised these stones in memory of Åslak and in memory of Göte, his son, and Åslak was Sven's brother."

Carduus acaulis, calyce glabro. Fl. Svec. 676 [*Cirsium acaule* Scop.] has only been seen in a few places before (*It. Oel. 157 & It. Gott. jul. 19*): [see entry of 25 June]. It was common on the ground between Omberg and Ösjö.

Omberg mountain half a mile from Ösjö was royal hunting ground; here we left the main road to see the top of the mountain, whence we had a marvellous view to the east and the south over vast cornfields and villages. On the southern slope of the Omberg were some beeches. On the western side were much spruce and a little *Taxus* or yew, and gooseberry bushes with smooth yellow berries. To the north, quite a distance from the summit, we saw the Apostle's Tree, a huge beech tree with eleven big trunks separated at the root, undoubtedly grown from an equal number of beech-nuts; some of them had fused slightly, and one of the middle ones had withered; one could not see any remains of the twelfth stem; thus I do not know whether the story is true that a farmer cut out the twelfth tree, alleging that Christ had only eleven apostles, since Judas had hanged himself.

One saw stalactites in all rills running down the Omberg; these were stone incrustations which had covered dry twigs fallen into the water, and are quite rare in Sweden. Below the main road one saw white bog lime here and there as an obvious indication that the water contains lime; thus it is easy to perceive the origin of the stalactites.

The gamekeeper had made a salt-lick out of salt mixed with clay and sand and put it here and there in square wooden boxes for the red deer and fallow deer to lick, just like farmers do for the cattle to make them thrive better. The farmer in Dalarna makes this salt-lick out of salt and bark with which he attracts the cattle and causes them to follow him wherever he goes in the woods.

We saw a delightful autumn picture between Nyby and Vadstena in the beautiful fields of Östergötland, where the farmers were reaping the fine wheat; some were harvesting peas and others mowed the pale corn. The sheaves were stooked in rows with ten pairs of sheaves leaning against each other in each stook. The crops were brought into the barn on wagons drawn by pairs of oxen. The geese collected the ears left by the reapers. The knotgrass was red and ripe in the fields as well as on the roads.

This night we rested in Vadstena after 5½ miles' journey.

August 23 [p. 338]

Vadstena town was not very large but attractive with its castle, church and the Soldier's Home.

The Soldier's Home is well built and fairly big and was formerly a monastery but is now a hostel for crippled soldiers.

The monastery church had many antiquities, which will be passed over since the late Archbishop E. Benzelius has described them; here we saw a picture of St. Britta [Saint Bridget of Sweden], a copy of the one we saw at Visingsö; in the graveyard there was an epitaph to a physician with the following inscription:

Hoc, ego qui recubo, duro sub fragmine faxi
multis dum vixi dulcis amicus eram,
aegris auxilium medicando ferre solebam
dum fuit humanis usibus apta manus,
Celsia nunc tracto, terrestria pharmaca linquens,
et Medico summo laetor adesse Deo.

In the choir lay Duke Magnus [d. 1596, the third son of Gustavus Vasa] carved life-size. In the chancel we saw the skulls of St. Britta and her daughter St. Cathrine; on the first skull the sutures were fused, indicating that she was old, but the skull was also polished, which told us that it had not been rotting in the earth but is likely to be *quid pro quo,* as the apothecaries say. St. Cathrine's sutures were still open except the *sutura coronalis,* which was closed.

Leontodon calyce toto erecto hispide foliis hispidis dentatis integerrimis Fl. Svec. 628 [*L.* 628 *asperum* of Index = *L. hispidum* L.] grew wild in the monastery garden. We saw it outside the town too, in the meadows. It has not been seen in Sweden before.

St. Britta's leek, of which the remarkable effect on moles has been much talked about in Stockholm and in the Royal Academy of Sciences, was the main reason why I returned home by way of Vadstena, for I wondered what this marvellous leek might be. It is well known here and its local name was "Munk-look" [Monk's leek]; it grew profusely in the monastery garden and was thought to be brought there from Italy by St. Britta. But I was much taken aback, as I expected to see a very rare plant, to find the Gotland ramsons, which I described in *It. Gott. 169* [entry of 25 June] or *Allium foliis lanceolatis scapo nudo semicylindraceo bulbo setis obvallato. Fl. Svec. 263.* [*A. ursinum* L.] *Sic minuit praesentia famam.*

The castle and half of the abbey church were said to be built of stone from Omberg, but the many Öland spikes [fossils] in the stone made it more likely that the stone was brought there from Öland; for, although there are like fossils in other Swedish bedrocks, they are seldom so numerous. In the castle, in Duke Magnus's rooms we saw a Diana with the inscription *Silvarum cultrix castissima virgo Diana est* and an Ariadne with the inscription *Mortalis Baccho placuit Ariadne marito.*

Sparganium [*S. erectum* L.] grew in the moat surrounding the castle, bigger than in other places in Sweden. It was as tall as a man; the leaves were two-fingers broad with a sharp edge on the back.

Superstitions provide more amusement than any use. Here they told us about one which is almost too ridiculous, namely that, if a knife which has been used for slaughtering swine and has remained unused thereafter is thrust into the earth in such a way that the farmer moves towards the knife all the time he is sowing, and the knife is moved to the other end of the field when he changes direction, no swine will ever dare to run into the field, for as soon as the swine reaches the edge of the field it will turn and run away as if it were actually stabbed with the knife. Anyone who has the stomach and constitution to believe this may do so.

We left Vadstena in the afternoon.

Eel-fishing was practised in the Motala river, with an eel-trap where it passes Motala inn on its way from Lake Vättern. This eel-trap was like a pipe with small slits in the bottom and enlarged at one end, closed with a grating, so that when the eel pushes itself in it cannot get out. Salmon-fishing was also practised in the Motala river by the Motala inn. Here was a 4-cornered dam and in this a 4-cornered cage with two chambers in it; these could be closed with a partition so that the salmon, entering with the stream, could be caught with a bag-net as soon as noticed by the watchers constantly on the look-out.

We found very deep sand on the road from Motala to Nykyrka.

The beautiful and well known mineral springs of Medevi were not far away; we rode there partly to taste the water, partly to visit the Chamberlain, Mr Carl de Geer, and see what insects he had caught this summer, for he has widely surpassed all native as well as foreign investigators in this branch of science.

Medevi mineral springs were situated in a lovely place; here were whole streets of houses for the guests who partook of the water; here were beautiful avenues and an excellent mineral water. This is the oldest spa in Sweden and undoubtedly still the best, although the spring has changed its outflow slightly. Much could be said about this water, enlarging on a *Spiritus*, a *Sulphur* or some other subterranean chemical principle, but we will leave the ancient doctrines and note that the water is impregnated with iron from an earth very similar to bog iron, which could in all likelihood be found south of Medevi if anyone would take the trouble to dig for it.

We remained in Medevi during the night.

August 24 [p. 341]

We continued our journey towards Stockholm.

NÄRKE

Hammar church kept us a little on the way, since we talked to the studious and inquisitive vicar, Mr [Daniel] Tiselius, who is well known for his treatises on Lake Vättern. We also saw a loon here.

The loon (*Animal. Svec. 121*) [*Colymbus arcticus* L., black-throated diver] was white below, black above; the back had white oblique lines and white spots on the wings. The sides of the throat and breast had wave-like white and black longitudinal lines. The upper parts of head and neck were grey; there was a purple long and broad patch from the bill to the breast and a white ring around the throat.

Various uncommon plants were found in Hammar, such as *Elatine* [*E. hydropiper* L.], *Limosella* [*L. aquatica* L.], *Hydrocharis* [*H. morsus-ranae* L.], *Utricularia nectario conico* [*U. vulgaris* L.], *Cicuta* [*C. virosa* L.] and both *Bidens* species [*B. tripartita* L. and *B. cernua* L.].

We rode through Askersund, a little town on the northern shore of Vättern. We had thus followed the shore of Vättern from its south end in Jönköping, along the eastern shore through Gränna and Vadstena until it ended here in Askersund.

We found an insect, or rather its grub, on both sides of the sandy road at some distance from Askersund. There were several little cylindric holes as big as if one had stuck a goose quill in the soil. They were half an ell [1 ft, 30 cm] deep and perpendicular; in the bottom was a hideous grub. The belly was oblong and pale; on the back it had a hump like a camel but two-cleft. The head was quite big and strong and seemed to consist of some tough horn-like matter, with two big black cruel teeth; the grub lay there on the bottom like a lion waiting for other insects to fall down into the den and be devoured by it. It is not known what kind of metamorphosis it undergoes or what kind of bird [sic!] it turns into.

We spent the night in Skyllberga inn.

August 25 [p. 342]

We passed through Örebro, a big and beautiful town; here we saw the castle and a tall *Populus nigra* in the castle garden; the District Governor, Baron Nils Reuterholm, told us about many interesting economic and physical features of this district; two of my travelling companions, J. Moraeus and H. J. Gahn, bid farewell to us here and rode towards the sulphur works at Dylta and the new copper mine at Falun.

VÄSTMANLAND

We hurried on, day and night.

We passed Reglan and Arboga in the night.

August 26 [p. 343]

The spring at Strömsholm was remarkable. It can be seen at a depth of 5 ells [10 ft, 3 m] on the bottom of the lake; when one rows over this, one almost shivers, because the blue bottom seems to be infinitely far away; all the plants looked blue although they are in fact green, such as *Confervae, Hottonia* [*H. palustris* L.] etc.; around it grew *Sagittaria* [*S. sagittifolia* L.]; *Dortmannia* [*Lobelia dortmanna* L.] and *Subularia* [*S. aquatica* L.] in abundance.

The water in this spring is incredibly clear, pure and cold, and clearer water than this can hardly be found anywhere; it surges up in a wide stream and drives the other water to the side. Although this water is so beautiful and clear, it is not very good for drinking, and I would not advise anyone to drink it daily.

There were two runic stones along the road about two miles from Västerås.

We saw the ruins of an old monastery on our right hand.

In Västerås we saw a mistletoe [*Viscum album* L.], brought there from Fullerö, the beautiful estate of the Senator, Count J. Cronstedt.

We spent this night in Säva bro.

UPPLAND

August 27 [p. 344]

We rode rapidly through Enköping to Uppsala.

In Uppsala, the old town and former capital of Sweden, I stayed a whole day. The incomparably well situated castle was still in ruins. The great church with its many royal epitaphs was visited; here was the Archbishop's residence, the District Governor's residence and the oldest University in Sweden; here is an excellent library, the fine stables had recently been repaired. The University, prompted by the learned professor of astronomy A. Celsius, had recently built a magnificent observatory. *Nosocomium Academicum* [University Hospital] and *Hortus Botanicus* [Botanic Garden] had not yet been salvaged. His Royal Majesty had shortly before my departure honoured me with a chair among the other professors in the medical Faculty.

August 28 [p. 344]

In the morning, our journey took us from Uppsala to Stockholm, where we arrived in the evening.

EXPLANATION OF PLATES

PLATE 1

Linnaeus's passport for his journey to Öland and Gotland in 1741.

PLATE 2

Öland journal for 6 June 1741, with sketch by Linnaeus of *Ranunculus illyricus* L.; see p. 98.

PLATE 3

Öland journal for 18 June 1741, with sketch by Linnaeus of pygidium of a trilobite, *Megalaspis* sp. (*Öländ. Resa*: p. 147), and of *Coccionella oblongo-guttata* L. (*Neomysia oblongo-guttata*); see p. 14.

PLATE 4

Gotland journal for 23 June 1741, with sketch by Linnaeus of wall and towers of Visby; see p. 109.

PLATE 5

Gotland journal for 23 June 1741, with sketch by Linnaeus of Valdemar's Cross at Visby commemorating those killed in a battle between Gotlanders and Danes in 1361; see p. 110.

PLATE 6

Gotland journal; sketches by Linnaeus of net (*Öländ. Resa:* p. 185) and harpoon (*Öländ. Resa:* p. 186) for seal-hunting; see p. 120.

PLATE 7

Gotland journal for 1 July 1741, with sketch by Linnaeus of weathered limestone blocks, the "Stone Giants", at Kyllej (*Öländ. Resa*: p. 218); see p. 136.

PLATE 8

Gotland journal for 6 July 1741, with sketch by Linnaeus of bench seat with reversible back (*Öländ. Resa:* p. 244); see p. 150.

PLATE 9

Artemisia rupestris L., 1753 (*A. foliis pinnatis, ramis adscendentibus* L., 1746) from Stora Karlsö (*Öländ. Resa*: facing p. 284); see p. 171.

Note. The text in Plates 2 to 5, 7 and 8 was written to Linnaeus's dictation by one of his companions (see p. 5), not by Linnaeus himself, and is not identical with the text printed later and here translated.

These plates are reduced from the original manuscript, which is 21 x 34 cm. This manuscript is housed in the Library of the Linnean Society of London, Burlington House, Piccadilly, London.

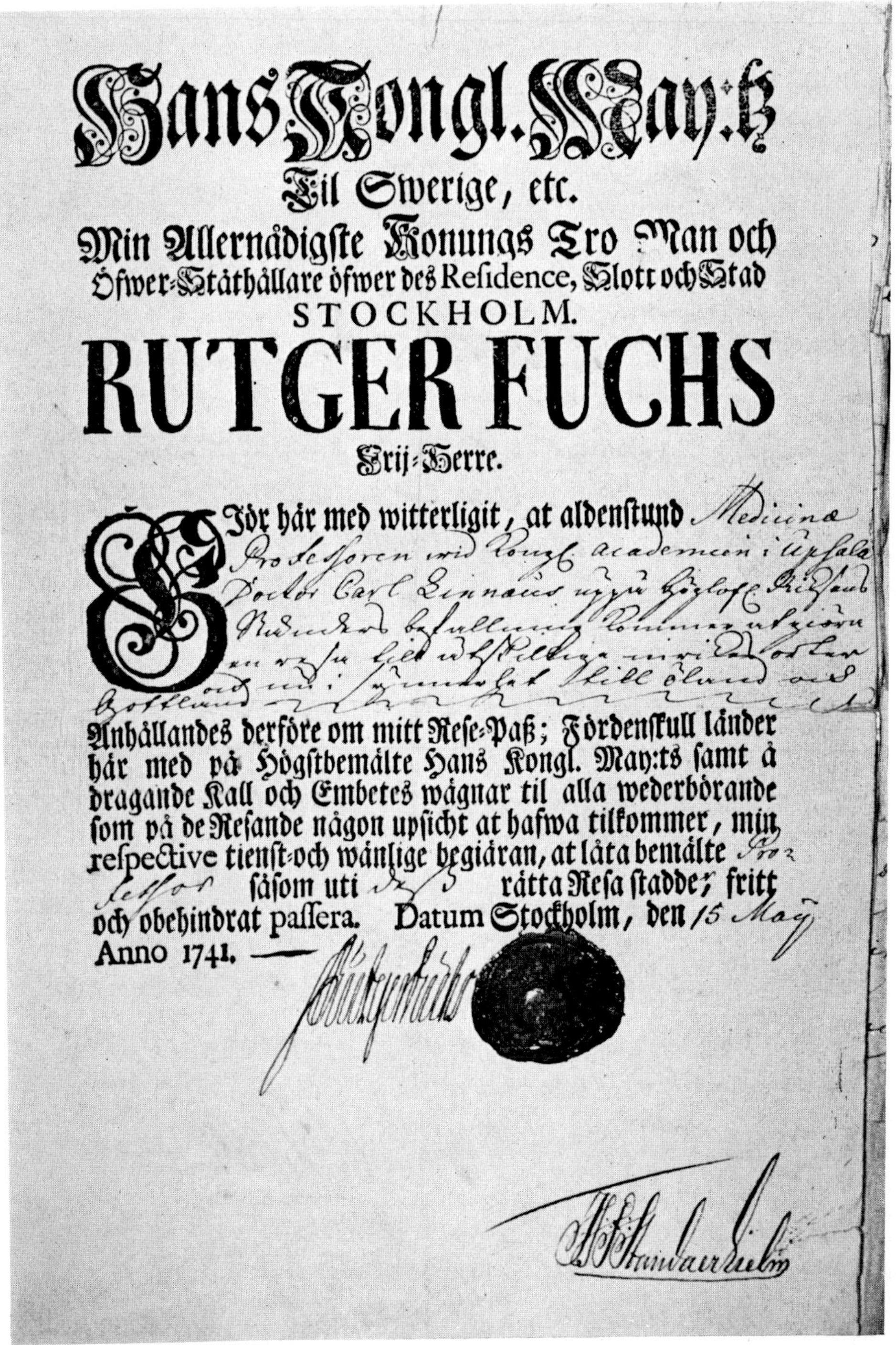

Hans Kongl. May:tz
Til Swerige, etc.
Min Allernådigste Konungs Tro Man och
Öfwer-Ståthållare öfwer des Residence, Slott och Stad
STOCKHOLM.
RUTGER FUCHS
Frij-Herre.

Gjör här med witterligit, at aldenstund Medicinæ Professoren wid Kongl. academien i Upsala Doctor Carl Linnæus uppå Högloflige Rikssens Ständers befallning kommer at giöra en resa till åtskilliga orter och nu i synnerhet till Öland och Gottland

Anhållandes derföre om mitt Rese-Paß; Fördenskull länder här med på Högstbemälte Hans Kongl. May:ts samt å dragande Kall och Embetes wägnar til alla wederbörande som på de Resande någon upsicht at hafwa tilkommer, min respective tienst- och wänlige begiäran, at låta bemälte Professor såsom uti des rätta Resa stadde; fritt och obehindrat passera. Datum Stockholm, den 15 May Anno 1741.

Junius 6 110

skar, Danskar och Tyskar täflat om denna
staden, hwilk wäl fastare bygnad [illegible]
ty hwad står emot krigs lagen ?
Staden låg in wid Stranden på wästra sidan
af landet utsträckt ~~som~~ innom en half circkel
inåt landet upför sidan af Berget, [illegible],
instängd med hög Mur med åtskillige [illegible]
[illegible] torn på gammalt maner med dubble
[illegible] doch förfallen, omgifne: Gatorne
woro gemena smala och trånga: Gränder-
ne helt irregulaire, husen tyske, somlige
af sten somliga af korsverk somliga af träd.
Taken mäst af tegel [illegible] från Tyskland
[illegible]
[illegible]
[illegible]
folket talade nästan Norska, på accenten woro
[illegible]: [illegible] sig om at wara [illegible].
Cicuta major, Echium majus, Cichoreum
~~[illegible]~~ Chenopodium folio obsolete-triangulo upsa-
liense, Anthriscus, Nasturtium Oelandicum
Bursa pastoris folio et Hyoseris minima
som tilförene i Swerige ej blifwit observe-
rad [illegible] på gator och wid kyrkorne.
De [illegible] som [illegible]
uti [illegible], woro [illegible]. Den
fi

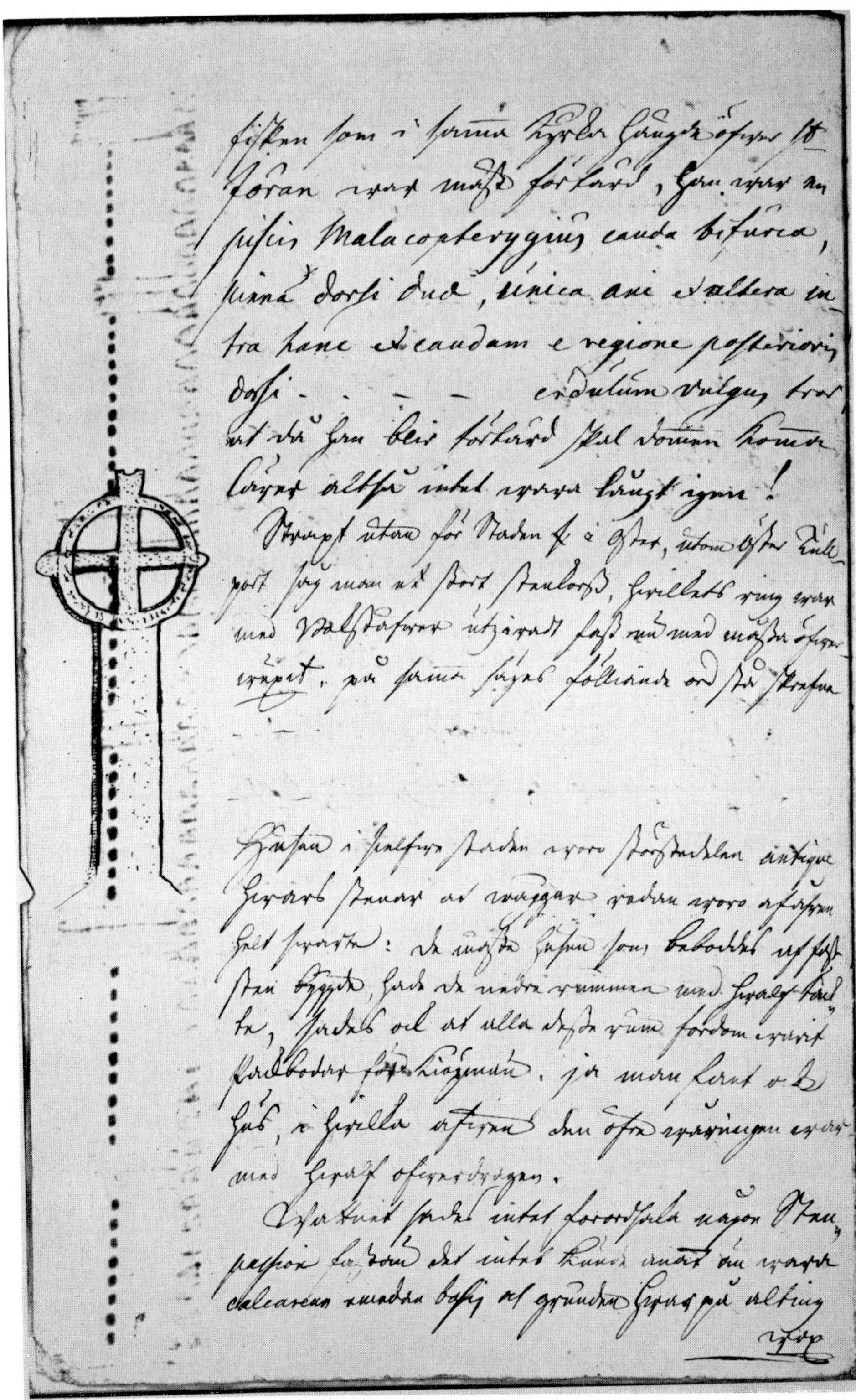

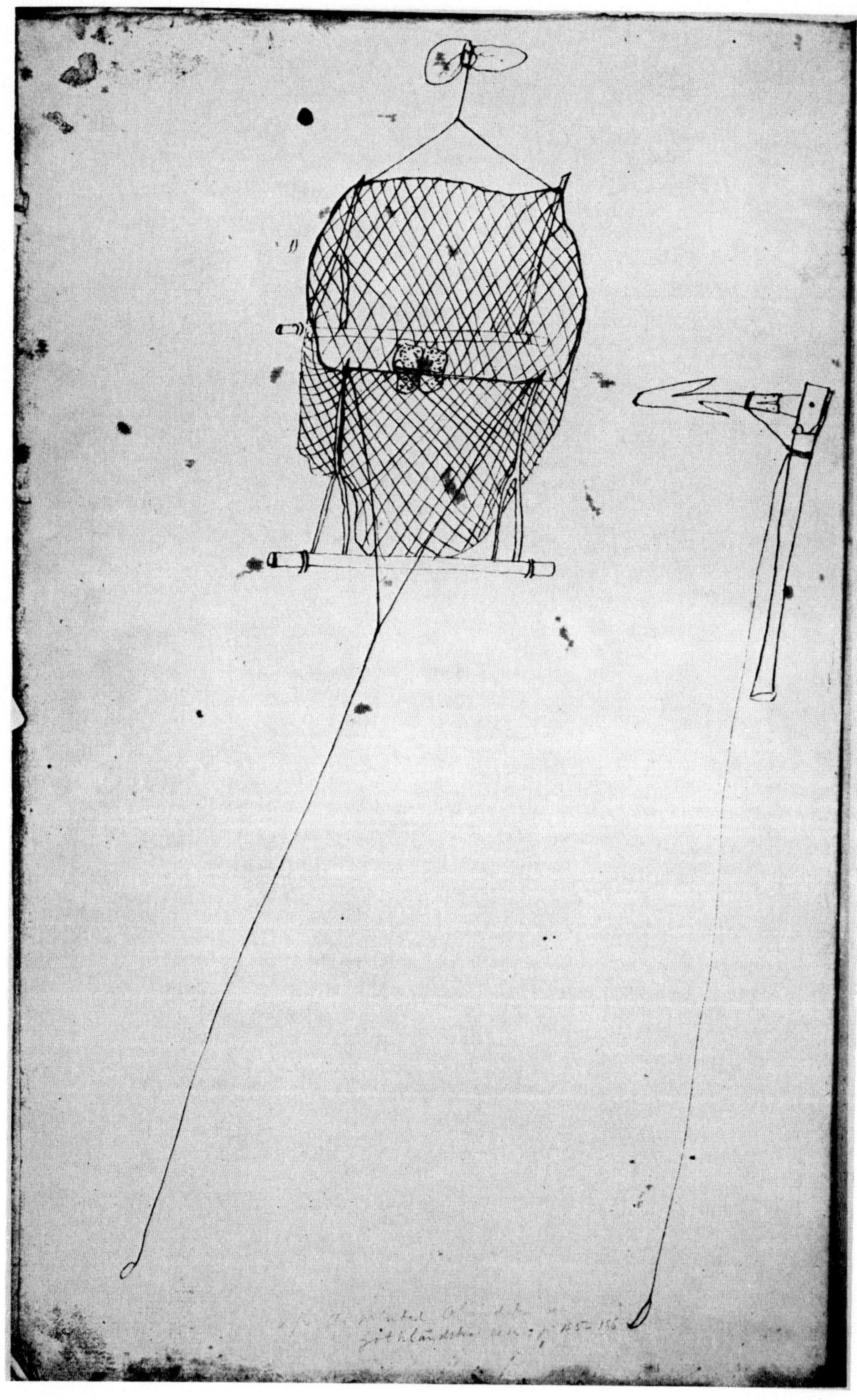

M. ÅSBERG AND W. T. STEARN

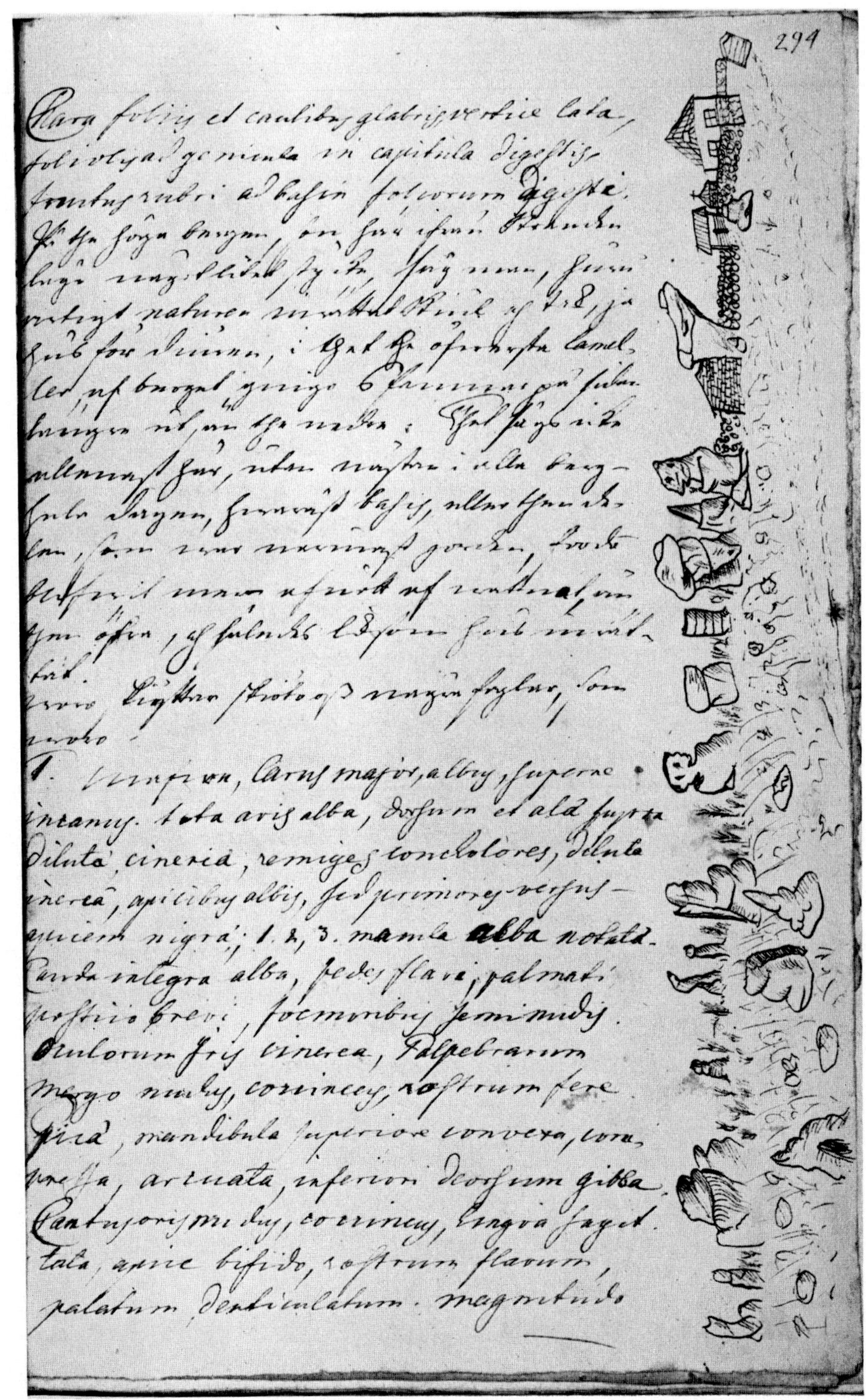

M. ÅSBERG AND W. T. STEARN

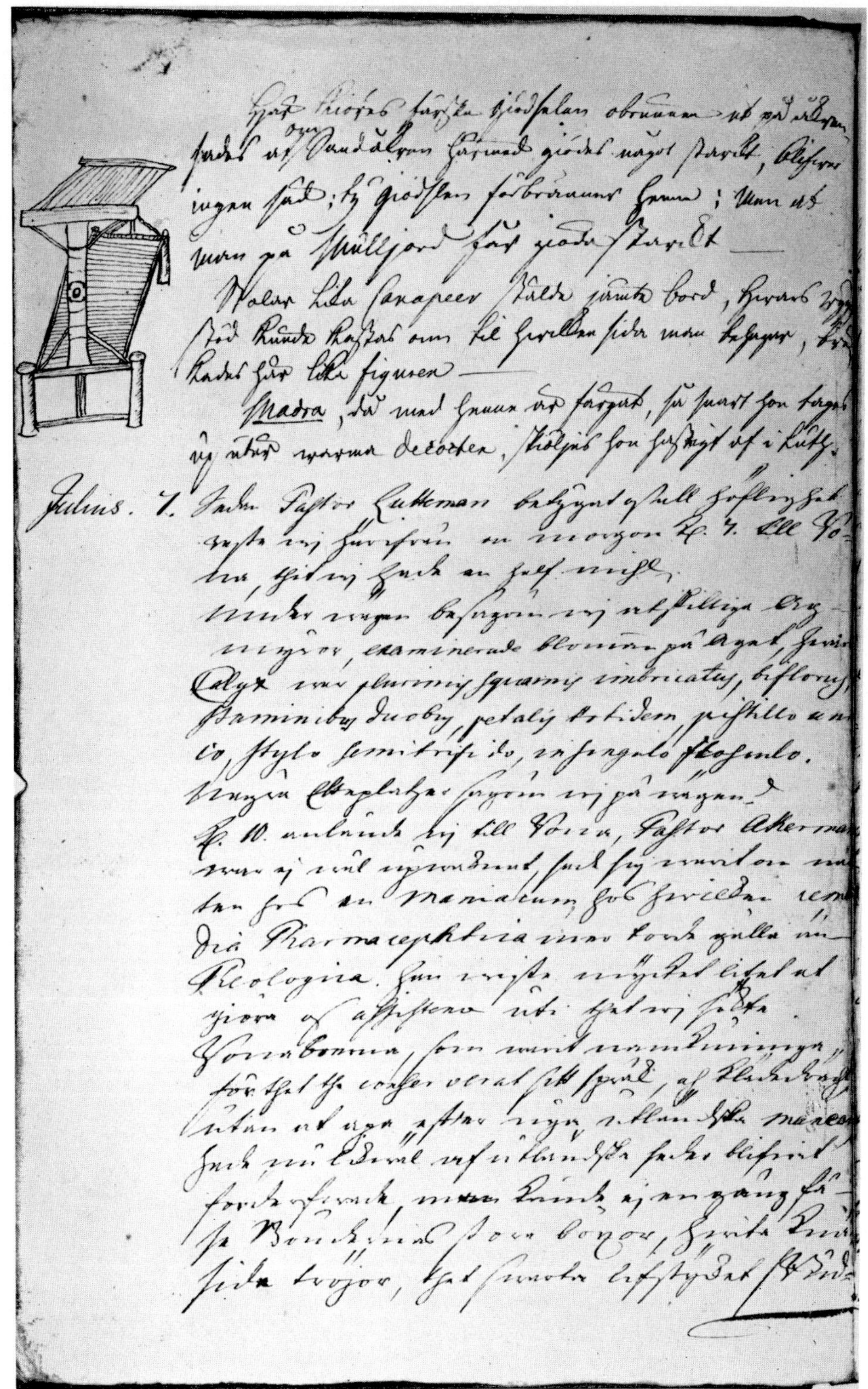

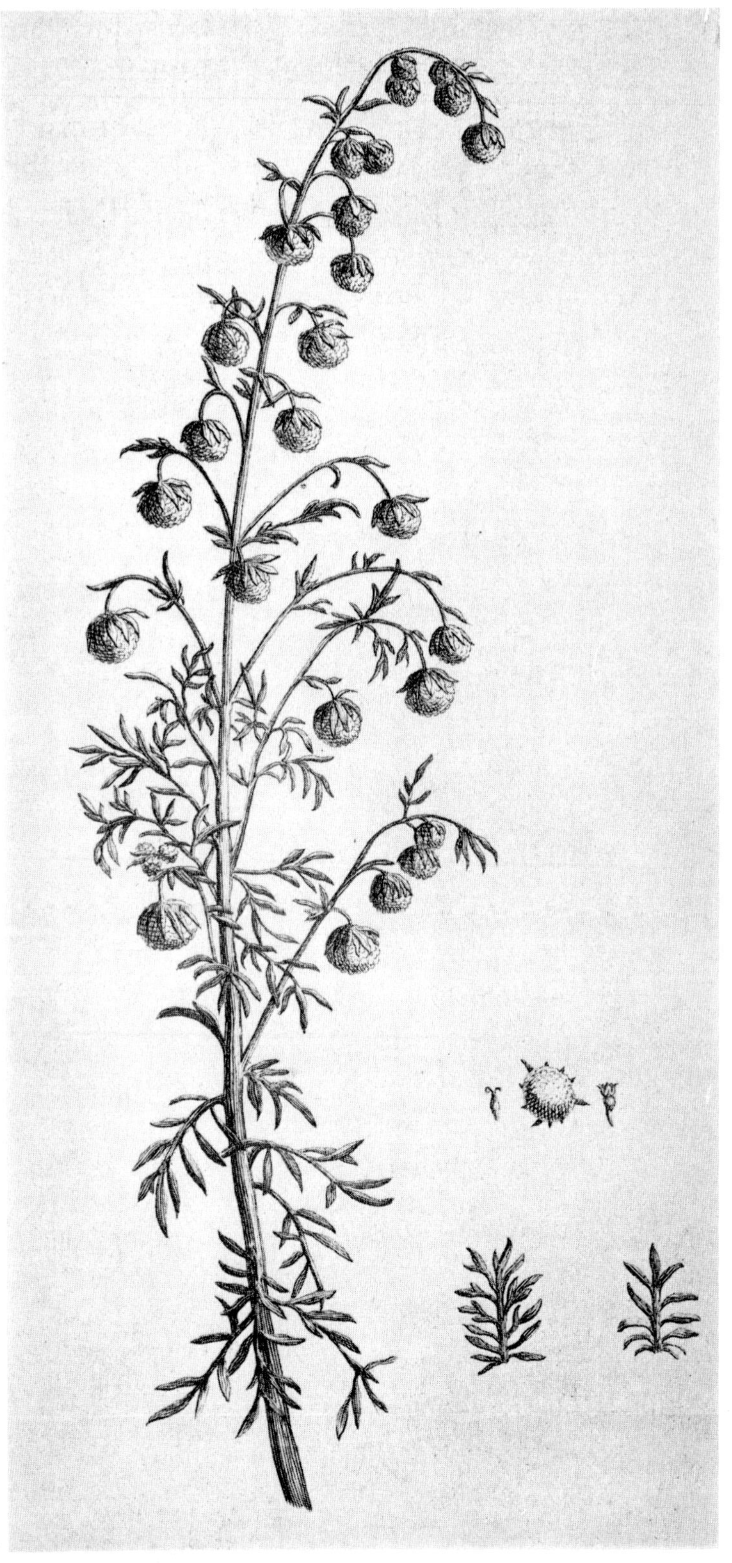

M. ÅSBERG AND W. T. STEARN

Map of Linnaeus's journey and facsimile of the botanical index to *Öländska och Gothländska Resa, 1745,* with binomial nomenclature for species (see p. 6); the numbers between the generic name and the species epithet (nomen triviale) refer to the numbered entries in Linnaeus's *Flora Suecica* (1745).

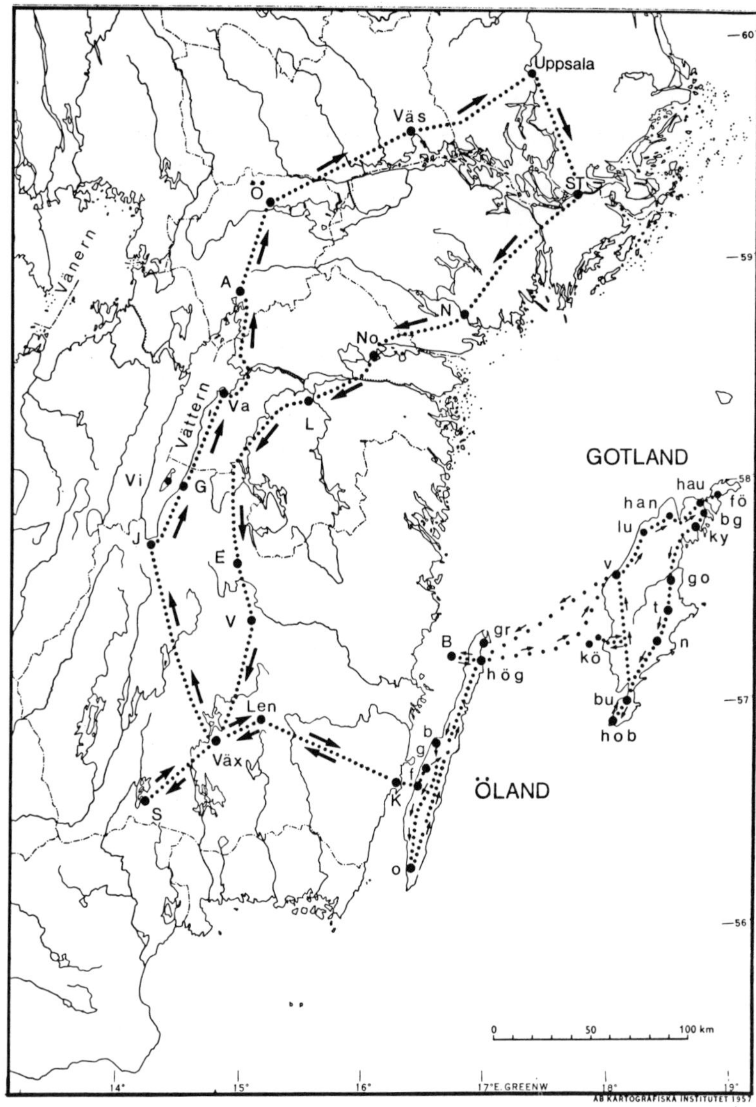

Linnaeus's journey to and from Öland and Gotland in 1741.

A, Askersund; B, Blåkulla (Jungfrun); b, Bornholm; bg, Bunge; bu, Burgsvik (Bursvik); E, Ekesjö; f, Färjestaden; Fo, Fårö; G, Gränna; g, Glomminge; go, Gothum; gr, Grankull; han, Hangvar; hau, Hau; hob, Hoburgen; hög, Högby & Böda; J, Jönköping; K, Kalmar; kö, Karlsöarna; ky, Kyllej; L, Linköping; Len, Lenhovda; lu, Lummelanda; N, Nyköping n, När; No, Norrköping; o, Ottenby; Ö, Örebro; S, Stenbrohult; ST, Stockholm; t, Torsburgen (Torsborg); V, Vetlanda (Vitlanda); Va, Vadstena; Väs Västerås; Väx, Växjo; Vi, Visingsö.

Wäxter på reſan ſedde.

MONANDRIA.

DIANDRIA.

Utri-

Aspe-

✱)o(✱

Aſperula 114. *odorata*. 60
239. 131.
115. *Madra*. 47.
72. 115. 168.
283. 187. 195
227. 238.
Galium 116. *luteum*. 39.
72. 96. 131.
228.
118. *album*. 39.
228. 283. 20.
119. *cruciata*. 127.
131.
Aparine 120. *ſcandens*. 131.
121. *pariſienſe*. 39
131.
Plantago 124. *lanceolata*.
20. 37. 131.
125. *lanata*. 216
214.
127. *anguina* 109
143. 233
128. *monanthos*.
316.
Sanguiſorba 130. *rubra*.
232.
Cornus 131. *Oſſea* 56. 72.
105. 291. 126.
235.
132. *herbacea*. 216.
Evonymus 133. *vulgaris*.
126. 47.
Alchemilla 135. *officinar*.
39. 62.
20.

Aphanes 137. *unica*. 316.
Potamogeton 139. *natans*
269.
146. *bocconis* 221

PENTANDRIA.

Myoſotis 149. *arvenſis* 130
aquatica 235
150. *Lappula* 274
Lithoſpermum 151. *officinar*.
70. 156
289.
Anchuſa 153. *Bugloſſum* 39.
Cynogloſſum 154. *vulgare*
61.
Pulmonaria 156. *immaculata* 81.
Echium 158. *vulgare* 164.
Aſperugo 159. *unica*. 127.
Androſace 160. *noſtras*. 56
57. 72.
Primula 161. *lutea* 39. 20.
237. 14.
162. *minor*. 39. 61
65. 146. 237.
38.
Menyanthes 163. *Trifolia*
18.
Hottonia 164. *unica*. 14.
32. 36.
Anagallis 169. *rubra*. 174
272.
Con-

Acha-

Saxi-

DODECANDRIA.

JCOSANDRIA.

POLYANDRIA.

Tha-

✱)o(✱

Den-

Hypericum 625. *Perforata.* 106. 131. 239. 253.

SYNGENESIA.

Leontodon 627. *Taraxac.* 1. 62. 72. 628. *Asperum* 629. *autumnale* 325

Hyoseris 631 *minima* 165.

Hieracium 633. *Pilosella.* 20. 72. 228. 637. *murorum.* 20.

Crepis 640 *tectorum* 60.

Prenanthes 645 *muralis.* 57. 215.

Lactuca 646. *quercifolia* 289.

Scorzonera 647 *humilis* 47. 61. 115. 36.

Tragopogon 648 *pratense* 334.

Lapsana 649 *vulgaris* 60. 156.

Cichorium 650 *agreste.* 61. 77. 101. 127. 164. 225. 247. 253.

Arctium 651 *Lappa* 47.

Carlina 652 *ramosa* 72. 274.

Onopordum. 653. *acanthium* 101. 269. 334

Carduus 654 *lanceolatus* 136. 655 *nutans* 157. 292. 293. 656 *acaulis* 157. 659 *palustris* 61.

Serratula 660 *nemorensis.* 47. 124. 38 662 *arvensis.* 78. 127.

Bidens 663 *tripartita* 98. 664 *integrifolia* 334

Tanacetum 666 *vulgare* 56 131.

Artemisia 668 *procumbens.* 53. 72 128. 669. *rarissima* 285. 670 *absinthium* 2 72. 98 101. 671 *Seriphium* 112. 250. 272.

Gnaphalium 672 *dioicum* 39. 20. 26. 674 *stoechas* 160

Doronicum 684. *Arnica.* 305. 316

Solidago 685. *Virgaurea.* 130. 283.

Se-

GYNANDRIA.

 Ophrys

✱)o(✱

Taxus

Aga-

Öländska och Gotländska namn på örter.

Wäxter 116 först i Sverige anmärkte

ÖLANDSKE.

449 Ane-

Index of animals, places and plants

Organisms for which the restricted type-locality (see p. 7) is or appears to be Öland are indicated by *, e.g. *Globularia vulgaris**; Gotland by **, e.g. *Geranium lucidum***; mainland Sweden by ***, e.g. *Apis tumulorum****. Those merely listed, e.g. on pp. 33, 43, 84, 91, 92, 110, 144, are for the most part not indexed below.

Currently used names have been added in parenthesis to Linnaean zoological binomials when different, e.g. *Coluber natrix (Natrix natrix), Chrysomela tanaceti (Galeruca tanaceti)*; a few of those adopted on pp. 13-107 have been amended; the co-operation of the zoological staff of the British Museum (Natural History) is here gratefully acknowledged. Obsolete spelling of place-names used by Linnaeus is given in parenthesis, e.g. Växjö (Wexiö). Although *å, ä* and *ö* follow *z* in the Swedish alphabet, they have been treated as *a*, ignoring diacritic marks, in the alphabetical sequence below.

W.T.S.

Index of Miscellanea